D1552022

To Prof. Thomas Huang,
with warm regards
Ajit Singh
July '92

Optic Flow Computation: A Unified Perspective

Optic Flow Computation

A Unified Perspective

Ajit Singh

The Institute of Electrical and Electronics Engineers, Inc.

Optic Flow Computation: A Unified Perspective

Ajit Singh

1951-1991

IEEE Computer Society Press
Los Alamitos, California
Washington • Brussels • Tokyo

IEEE COMPUTER SOCIETY PRESS MONOGRAPH

Library of Congress Cataloging-in-Publication Data

Singh, Ajit, 1963-

Optic flow computation : a unified perspective / Ajit Singh.
p. cm.
Includes bibliographical references.
ISBN 0-8186-2602-X (case) -- ISBN 0-8186-2601-1 (m/f).
1. Image processing. 2. Motion Perception (Vision)--Mathematical models. I. Title
TA1632.S595 1991
621.36'7--dc20 91-32949
CIP

Published by the
IEEE Computer Society Press
10662 Los Vaqueros Circle
PO Box 3014
Los Alamitos, CA 90720-1264

IEEE Computer Society Press Order Number 2602
Library of Congress Number 91-32949
IEEE Catalog Number 91EH0343-4
ISBN 0-8186-2601-1 (microfiche)
ISBN 0-8186-2602-X (case)

Additional copies can be ordered from

IEEE Computer Society Press
Customer Service Center
10662 Los Vaqueros Circle
PO Box 3014
Los Alamitos, CA 90720-1264

IEEE Service Center
445 Hoes Lane
PO Box 1331
Piscataway, NJ 08855-1331

IEEE Computer Society
13, avenue de l'Aquilon
B-1200 Brussels
BELGIUM

IEEE Computer Society
Ooshima Building
2-19-1 Minami-Aoyama
Minato-ku, Tokyo 107
JAPAN

Cover design: Joe Daigle
Editorial production: Penny Storms
Printed in the United States of America by Braun-Brumfield, Inc.

THE INSTITUTE OF ELECTRICAL AND ELECTRONICS ENGINEERS, INC.

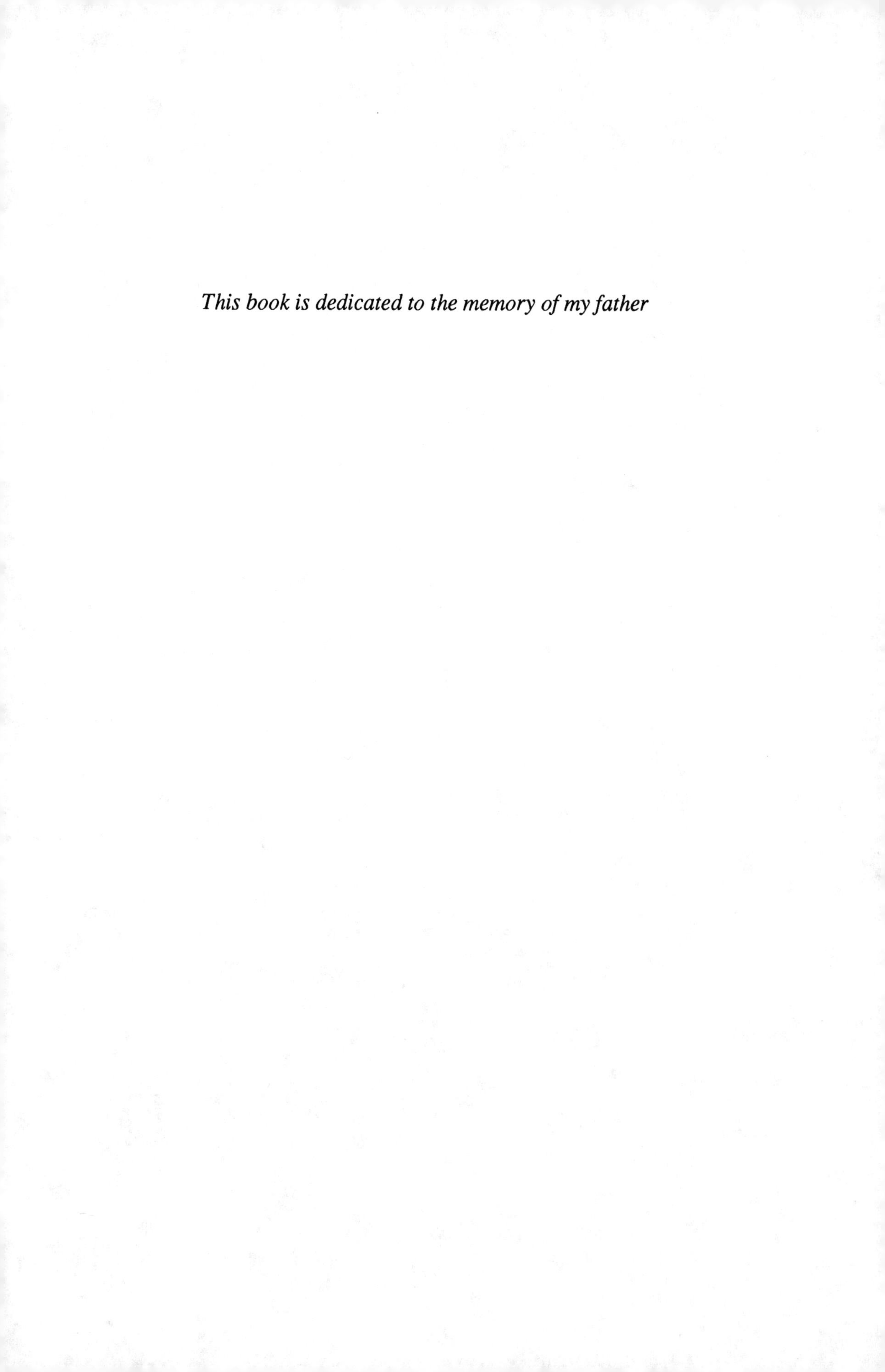

This book is dedicated to the memory of my father

Preface

The recent past of visual-motion research has witnessed a significant amount of work on optic-flow computation. Numerous frameworks have been proposed that use one of the three major approaches to recover optic flow—the gradient-based approach, the correlation-based approach, or the spatiotemporal energy-based approach—and one of several smoothing-based approaches to regularize it. This book is a two-pronged attempt to develop a new estimation-theoretic framework for optic-flow computation and to unify and integrate the existing approaches in view of this framework.

The new framework developed in this book views the problem of recovering optic flow from time-varying imagery as a parameter-estimation problem. It applies some existing techniques of statistical estimation theory to compute optic flow. In particular, this framework classifies the optic-flow information available in time-varying imagery into two categories: conservation information and neighborhood information. It recovers each type of information in the form of an *estimate* accompanied by a *covariance matrix*. To compute optic flow, it fuses the two estimates on the basis of their covariance matrices. This framework allows estimation of certain types of discontinuous flow fields without any a priori knowledge about the location of discontinuities. In other words, the flow fields estimated using this framework are not blurred at motion discontinuities. Also, this framework lends itself very naturally to recursive estimation of three-dimensional scene geometry from optic flow using Kalman-filtering-based techniques. Finally, this framework serves as a platform to unify the various existing approaches for optic-flow computation.

In the context of this book, unification has two aspects, one pertaining to conservation information and the other to neighborhood information. With respect to conservation information, I show that the new framework is applicable identically to each of the three major approaches. I show that a single procedure can recover the velocity information using any one of these approaches. With respect to neighborhood information, I show that the formulation used in this framework reduces to some of the existing smoothing-based formulations under various simplifying assumptions.

Unification serves several objectives. It provides a platform to compare the underlying assumptions and models of various frameworks that have appeared or will appear in the literature. In fact, it can be used to generate new approaches. Furthermore, it can be used to predict the performance of a given framework under a given set of operating conditions. Finally, it provides the basis for integration.

Integration, in the context of this book, is a computational paradigm where the aforementioned three approaches cooperate to give a robust estimate of optic flow. The motivation for integration is the following. Because the underlying measurements used by the gradient-based, correlation-based, and spatiotemporal energy-based approaches are different, they have different error characteristics. I illustrate this fact with examples in the course of this book. This scenario is representative of the classic multisensor problem. Algorithms based on the three approaches can be thought of as three different sensors measuring a given quantity—optic flow—with different error characteristics. The measurements from different sensors can be integrated to produce an estimate of optic flow that has the minimum mean-squared error. The confidence associated with this estimate is higher than the confidence associated with any one of the individual measurements. I apply the principles of statistical estimation theory to develop a framework for integration. I also show two algorithms based on this framework.

In addition to the major issues mentioned above, I address a number of related issues. First, I clarify the distinction between image flow and optic flow. The two quantities are related but not identical. The conditions under which the two quantities are equal are derived using basic principles of photometry.

Second, I review the past research on optic-flow computation from a new perspective. The problem of recovering optic flow from time-varying imagery underlies two functional steps. I analyze representative techniques from past research in view of this two-step solution. Next, I reformulate the *aperture problem.* In most past research, it has been regarded as a "binary" problem: Either it exists or it doesn't. This book formulates the aperture problem as a "continuous" one. It always exists, but to varying extents: from very negligible (in the vicinity of corners) to very acute (in the vicinity of very sharp edges). This is the formulation that unifies the three approaches.

Finally, I address the issue of incremental estimation of optic flow and its significance in real-time applications and derive a scheme for such incremental estimation from the framework outlined above.

This book is based on my doctoral dissertation at Columbia University. I wish to thank Peter Allen (Columbia University), Padmanabhan Anandan (Yale University), and Michael Shneier (Philips Laboratories) for motivating this work and providing invaluable advice and encouragement throughout the course of research. Many people have contributed directly or indirectly. I can only give a partial list. John Kender introduced me to the fascinating world of vision research and brought me to Columbia. Terry Boult and Martin Vetterli gave me the much required reviews of this book. Ernie Kent took me to the "real world" at Philips, which turned out to be the ideal world to do research. My colleagues at Columbia University, Philips, and Siemens—Tej Anand, Laura Appleton, Ming-Yee Chiu, Sandeep Dalal, Leo Dorst, Roy Featherstone, Davi Geiger, David Marimont, Inder Mandhyan, Thom Warmerdam, Larry Wolff, George Wolberg, and Hsiang-Lung Wu—were always there when I needed their advice and help. Surendra Ranganath helped me regain my literacy in mathematics. Sandeep Mehta provided numerous hours of help in building various display tools required for analyzing the experimental results. Henry Ayling, Jon Butler, Ez Nahourai, Penny Storms, and Rao

Vemuri gave the much needed support during the review and production of this book. Tom Culviner copyedited the manuscript and composed the pages. Vikas Joshi, whom I met during my brief sojourn at Syracuse University, my parents, and my brother have been a great source of inspiration.

Finally, there is a very amazing woman, Princy, who worked very hard on this book, even though she did not write it. She is married to me. Any shortcomings in this book are as much her fault as mine!

Table of Contents

Chapter 1

Introduction

Parmenides (ca. 450 B.C.) thought that the world was ever constant, "Everything is static." Heraclitus took the other extreme view, "The world is constantly in motion" [1]. The philosophical controversy over the static or dynamic nature of the world aside, the relationship between motion and perceived structure has intrigued the human mind throughout history. In *Optics*, Euclid (ca. 300 B.C.) wrote [2]

> In the case of flat surfaces lying below the level of eye, the more remote parts appear higher.... In the case of objects below the level of the eye which rise one above another, as the eye approaches the objects, the taller one appears to gain height, but as the eye recedes, the shorter appears to gain.... In the case of objects of unequal size above the eye which rise one above the other, as the eye approaches the objects, the shorter one appears to gain height, but, as the eye recedes, the taller one appears to gain.

A systematic study and a written record of the relationship between the three-dimensional structure of objects and their apparent motion dates back to the nineteenth century. Helmholtz studied this problem in detail. In his seminal work of 1866, *Physiological Optics*, he wrote [3]

> In walking along, the objects that are at rest by the wayside stay behind us; that is, they appear to glide past us in our field of view in the opposite direction to that in which we are advancing. More distant objects do the same way, only more slowly, while very remote bodies like the stars maintain their permanent positions in the field of view, provided the direction of the head and body keep their same directions. Evidently, under these circumstances, the apparent angular velocities in the field of view will be inversely proportional to their real distances; and, consequently, safe conclusions can be drawn as to the real distance of the body from its apparent angular velocity.

Since then, perception of motion has drawn the attention of researchers from a variety of disciplines including psychology, physiology, and computational vision. This research has a dual objective: to understand how humans perceive motion and to formulate a computational theory for motion perception that could be used in an automaton. During this phase of research, the term "optic flow" came into being. This book is about optic flow. Before I discuss what optic flow is, what it is useful for, and, most importantly for this book, how it is estimated, I give a brief review of the general area of visual motion perception.

Visual motion perception: An overview

Perception of motion is an important function of the human visual system. It plays a central role in the perception of depth, segregation of objects, and estimation of the three-dimensional motion of objects in the visible world. Numerous classifications have been proposed for the plethora of research in this field. I prefer to classify the literature into two categories.

Two categories of research. In the first category there are studies that discuss the various effects associated with motion perception. In this category are some of the earlier works in psychology, such as Miles's study [4] of the apparent reversed motion of a fan and the work of Wallach and O'Connell [5] on kinetic depth effect, as well as Ullman's recent experiments [6] using two counter-rotating transparent cylinders.

The second category comprises numerous efforts to explain these effects and duplicate them on a computer. These efforts span at least three distinct standpoints: empirical, normative, and theoretical. Proponents of the empirical approach try to determine *what is* the mechanism by which humans or other animals perceive visual motion. This approach underlies the research in disciplines such as sensory physiology [7-17] and psychology [18-33]. The normative standpoint deals with the question of *what should be* the mechanism for visual motion perception (to perform a given task, say, navigation, in some optimal sense). Similarly, the theoretical standpoint is concerned with identifying *what could be* the various possible mechanisms for understanding motion. A significant amount of research in computational vision [6, 34-45] attempts to answer normative and theoretical questions. In addition, there have been efforts to estimate motion from the signal-processing perspective—for applications in communication and television technology [46, 47]. Since this book primarily studies the theoretical question of recovering a possible representation of visual motion, a thorough review of all this literature would be both inefficient and irrelevant. I will just identify two important issues and discuss them in the next section.

- Perception of motion could be thought of as a two-step process [6, 20]. The two steps are *measurement* of the two-dimensional motion projected on the imaging surface and *interpretation* of the 2D motion to draw conclusions about the 3D structure and motion of the scene. (Most researchers consider motion perception to be two-step process, but there isn't ample biological evidence to support this view. Furthermore, many researchers in computational vision have treated the two steps in an integrated fashion [48-50].)

- There could be (and possibly are) two distinct mechanisms for measurement of 2D projected motion in most advanced animals including humans: *short-range* mechanisms that result into an instantaneous characterization of image motion as a 2D velocity field, and *long-range* mechanisms that underlie the "discrete snapshots" approach, where the salient features discovered in two discrete snapshots of the world are matched to recover the spatial displacement of each feature.

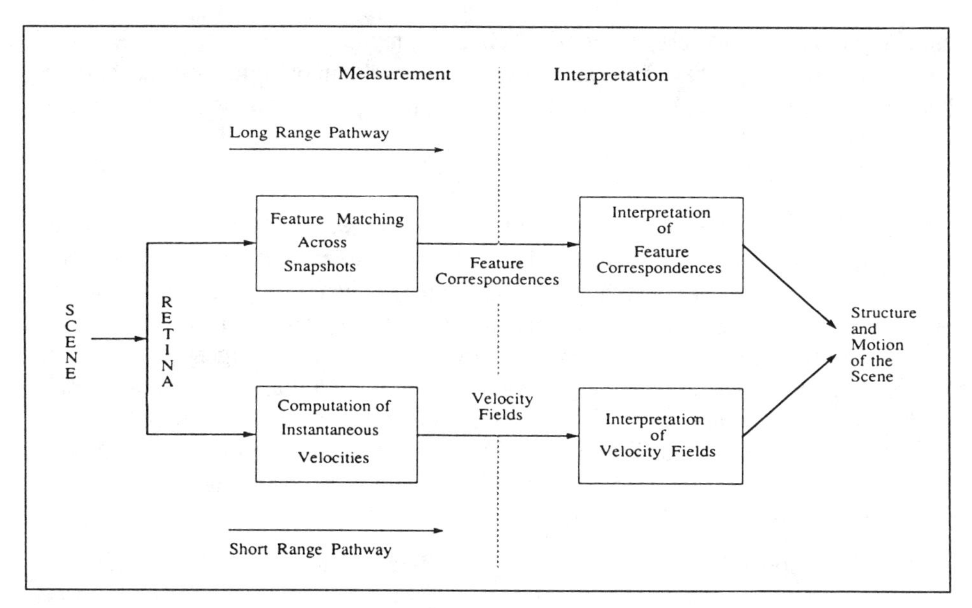

Figure 1.1. Motion perception as a twofold dichotomy.

Motion perception as a twofold dichotomy. Figure 1.1 gives a possible functional description of the process of motion perception based on the summary in the preceding section. A twofold dichotomy is apparent in this description. First, there are two possible mechanisms: long-range and short-range. Second, in each one of these mechanisms, there are two stages of perception: measurement and interpretation. A brief description of this twofold dichotomy follows.

Long-range mechanisms: Measurement and interpretation. Ullman's work [6] is a classic example of a long-range mechanism. Its underlying approach is also referred to as the "discrete snapshots" approach. The *measurement* stage in this approach is responsible for identifying distinctive features in two or more discrete snapshots of the dynamic scene and matching them across the snapshots. This is referred to as *token matching* and is analogous to the correspondence problem in stereopsis [51-52]. Thus, the output of the measurement stage can be visualized as a map that gives the position of various features (also called tokens) in a set of images taken at discrete instants of time.

The *interpretation* stage uses this map to derive the 3D positions of all the points that correspond to the tokens and velocities of the rigid bodies that contain these points. In essence, this approach provides 3D information for only a sparse set of points, rather than everywhere in the visual field.

Ullman discussed in detail the psychological and psychophysical aspects of the problem of token matching [6]. Since then, the computational issues related to token matching have been addressed extensively in the computer vision literature [6, 53-58]. Also, several solutions have been suggested for the problem of feature-correspondence interpretation (to derive 3D inference) under a variety of assumptions about the underlying geometry [55, 59-65].

Short-range mechanisms: Measurement and interpretation. The short-range mechanisms encompass what has been referred to as the image-flow approach. (I use the terms *image flow* and *optic flow* interchangeably, until I define them and explain the distinction between them in the next chapter.) In this approach, the measurement stage is responsible for constructing a 2D optic-flow field from time-varying imagery of the scene. Informally, optic flow can be thought of as a representation that has—associated with every point on the retinal field—a 2D velocity vector that depicts the projection of the 3D velocity of the corresponding point in the scene. The interpretation stage takes the velocity field as its input to extract information about the depth (or surface orientation) and the velocity of every point in the visual field.

The issue of optic-flow measurement has been addressed in the research in physiology, psychology, and computational vision alike. Work by Hubel and Wiesel [11, 12], Bridgeman [9], and Grusser and Grusser [10] typifies the physiological studies of neuronal mechanisms that support the computation of instantaneous 2D velocities. Recently, there has been a deluge of efforts in psychology and computational vision to estimate optic flow from time-varying imagery. These efforts can be classified into

- gradient-based approaches [34, 40, 66-79],
- correlation-based approaches [80-84], and
- spatiotemporal energy-based approaches [85-91].

Methods such as the phase-correlation methods or LMS recursive methods [46, 47], which span one or more of these categories, have been applied in communication and television theory. Also, there are methods that use closely spaced (in time) images to estimate motion, but they do not recover optic flow as an explicit representation of the 2D velocity field [92]. Such methods are not discussed in this book.

Work on interpretation of image flow started long before work on its measurement. J.J. Gibson, one of its earliest proponents, believed that image flow was the fundamental stimulus to the visual system [26]:

> Let us recall once again that the arrested optic array is an unusual case of the changing array; it is obtained in a frozen world by an observer who holds still and uses one eye. The eye continues to work but it is not what the organ evolved for. Optical rest is a special case of optical motion, not the other way around. The eye *developed* to register change and transformation. The retinal image is seldom an arrested image in life. Accordingly, we ought to treat the motion picture as the basic form of depiction and the painting or photograph as a special form of it. What a strange idea! It goes counter to all we have been told about optics. But it follows directly from ecological optics. Moviemakers are closer to life than picture makers.

Study of the mathematics of image-flow interpretation began soon after Gibson's original postulation, but only for special cases [23, 93, 94]. General treatments of the problem have appeared only recently [39, 44, 45, 95-102]. Even though the underlying issues are well understood, the controversy over whether or not optic flow is used in human vision goes on:

> The fundamental ecological input for vision is not a camera-like time-frozen image but a constantly changing optic array or flow field. —Lee [31]

> It remains to be seen whether optic flow is used in human vision. —Marr [51]

About this book

This book is about estimation of optic flow, one of the possible representations of visual motion. In light of the twofold dichotomy suggested earlier, this book is concerned with what goes on in the lower left box in Figure 1.1. The recent past of visual-motion research has seen much work in this area. Numerous frameworks have been proposed that make use of one of the three basic approaches: the gradient-based approach, the correlation-based approach, or the spatiotemporal energy-based approach. This book is a two-pronged attempt to

- develop a new estimation-theoretic framework for optic-flow estimation, and
- unify and integrate the existing approaches in view of this framework.

In Chapter 2, I review some basic definitions, clarify the distinction between image flow and optic flow, and derive the conditions under which the two quantities are equal. I also explain my preference for the term *estimation* instead of the more commonly used term *measurement*—even though I have used the two interchangeably so far. Finally, before proceeding with the main subject of this book, optic-flow estimation, I review some of the applications of optic flow: recovering 3D structure and motion, performing motion-compensated image-sequence enhancement, and so on. For readers uninitiated in optic flow, this review should provide the necessary motivation.

In Chapter 3, I review the state-of-the-art techniques for optic-flow estimation. I view optic-flow estimation as a two-step process:

1. Derive locally available velocity information in the form of conservation constraints.
2. Propagate velocity information using neighborhood constraints.

I classify the current approaches for the first step into three categories: gradient-based, correlation-based, and spatiotemporal energy-based approaches. Likewise, I classify the current approaches for the second step (which amounts to overcoming the well-known aperture problem) into two categories: approaches based on a smoothness constraint and approaches based on the analytic structure of optic flows.

Also in Chapter 3, I review some representative optic-flow estimation techniques from the past research in light of the preceding classifications and evaluate their performance. In doing so, I clarify exactly how much of the performance (or lack of it) is due to the first step and how much to the second step. This distinction is very important for designing robust estimation techniques. Unfortunately, it has been ignored in most previous reviews [80, 103, 104].

The next two chapters are devoted to a new framework for optic-flow estimation. What is new about this framework? It views the problem of recovering optic flow from time-varying imagery as a parameter-estimation problem and applies some of the techniques commonly used in estimation theory to recover optic flow. For this reason, I call this framework estimation-theoretic.

In Chapter 4, I discuss the general nature of motion information that can be derived from small spatiotemporal neighborhoods in time-varying imagery. I show that this information is of two types: conservation information and neighborhood information. Furthermore, both can be regarded as inexact.

In Chapter 5, I show the details of the framework. In essence, it fuses the two types of information statistically to recover optic flow. I discuss how to recover conservation information as well as neighborhood information from the imagery and describe a method to combine them. I also show two algorithms based on this framework. A very important issue related to optic flow is its incremental estimation over time for real-time applications. The framework developed in these two chapters can be extended very naturally to include this dimension. Strategies for such extension are presented in the appendices at the end of the book.

Chapters 6 and 7 are devoted to experiments. In Chapter 6, I use the two algorithms mentioned above to estimate optic flow from a variety of image sequences. Chapter 7, on the other hand, is devoted to applications of optic flow. I describe both 3D and 2D applications. For a 3D application, I show that the optic-flow fields recovered by the new framework can be used to obtain useful depth information via a Kalman-filtering-based framework. As an example of a 2D application, I use the optic-flow fields recovered by the framework to perform a motion-compensated enhancement of image sequences.

Chapter 8 is devoted to unification. There are two aspects of unification, pertaining to conservation information and neighborhood information, respectively. For conservation information, I show that the framework developed in the preceding three chapters is applicable to each of the three basic approaches. I show that a single procedure can be used to compute locally available velocity information using any one of these approaches. With respect to neighborhood information, I show that the formulation used in this framework reduces to some of the existing smoothing-based formulations under various simplifying assumptions.

I also show that unification serves several objectives. It provides a platform to compare the underlying assumptions and models of various frameworks that have appeared or will appear in the literature. In fact, it can be used to generate new approaches. Furthermore, it can be used to predict the performance of a given approach under a given set of operating conditions. Finally, it provides the basis for integration.

Chapter 9 is devoted to integration. I show that optic-flow estimates obtained by using any one of the three approaches mentioned above have different error characteristics.

This scenario is representative of the classic multisensor problem. Algorithms based on the three basic approaches can be thought of as three different sensors measuring a given quantity—optic flow—with different error characteristics. The measurements from different sensors can be combined to produce an estimate of optic flow that minimizes the mean-squared estimation error.

In other words, the three basic approaches can be integrated to give an estimate of optic flow that has a higher confidence than the estimate obtained from any one approach alone. I show that the framework developed in this book can serve as a platform for such integration. I apply principles of statistical estimation theory to develop a scheme for integration. I also show two algorithms based on this scheme.

The field of optic-flow estimation is so wide and the possibilities for extension of this work so many that it would be premature to give a conclusion. I do, however, make an attempt in Chapter 10.

Chapter 2

Image Flow and Optic Flow

I will start by defining image flow and optic flow, two terms often confused and used interchangeably. The difference between the two could be quite significant. I derive the conditions under which the two quantities are equal. I also explain my preference for the term *estimation* (of optic flow) rather than the more commonly used term *measurement*. Finally, I review some applications of optic flow—recovering 3D structure and motion, motion compensation, and so on. This review is not intended to be an exhaustive survey of the existing literature on the usefulness of optic flow, but to motivate readers unfamiliar with optic flow and set the stage for the main subject of this book: how to compute optic flow.

Definitions

An *image-flow field* depicts, at each point on the imaging surface, a 2D projection of the instantaneous 3D velocity of the corresponding point in the scene. The scenario in Figure 2.1 can be used to explain this definition. A rigid body B is undergoing an arbitrary motion relative to the imaging surface. A point P on the rigid body has a velocity $\mathbf{S}$ with respect to a *world coordinate system* (X, Y, Z) fixed at the origin O_W. The imaging surface, a plane in this case, is fixed with respect to the world coordinate system in such a way that O_W is the center of projection, or the viewpoint [105]. An *image coordinate system* (x, y) is attached to the image plane with its origin at O_i. The point p on the image plane is the projection of the scene point P. Likewise, the vector $\mathbf{V} = (u, v)$ in the image plane, originating at p, is the projection of the velocity vector $\mathbf{S}$. By the definition given above, the vector $\mathbf{V}$ is the image-flow vector at the point p. The distribution of such vectors over the entire imaging surface comprises the image-flow field. Figure 2.2 shows an example of a flow field.

Although I have used a planar imaging surface to illustrate image flow, it is by no means a part of the definition. The imaging surface could have any shape. In the case of the human eye, for example, the imaging surface is roughly a spherical frustum. Typical cameras, however, have a planar imaging surface. For the purpose of this book, I assume a planar imaging surface.

The *optic-flow field* is the 2D distribution of *apparent velocities* that can be associated with the variation of brightness patterns on the image [34]. There is no mention of the *scene* in this definition. The scene does not have to be in motion relative to the image for the optic-flow field to be nonzero. As I will show later in this chapter, optic flow

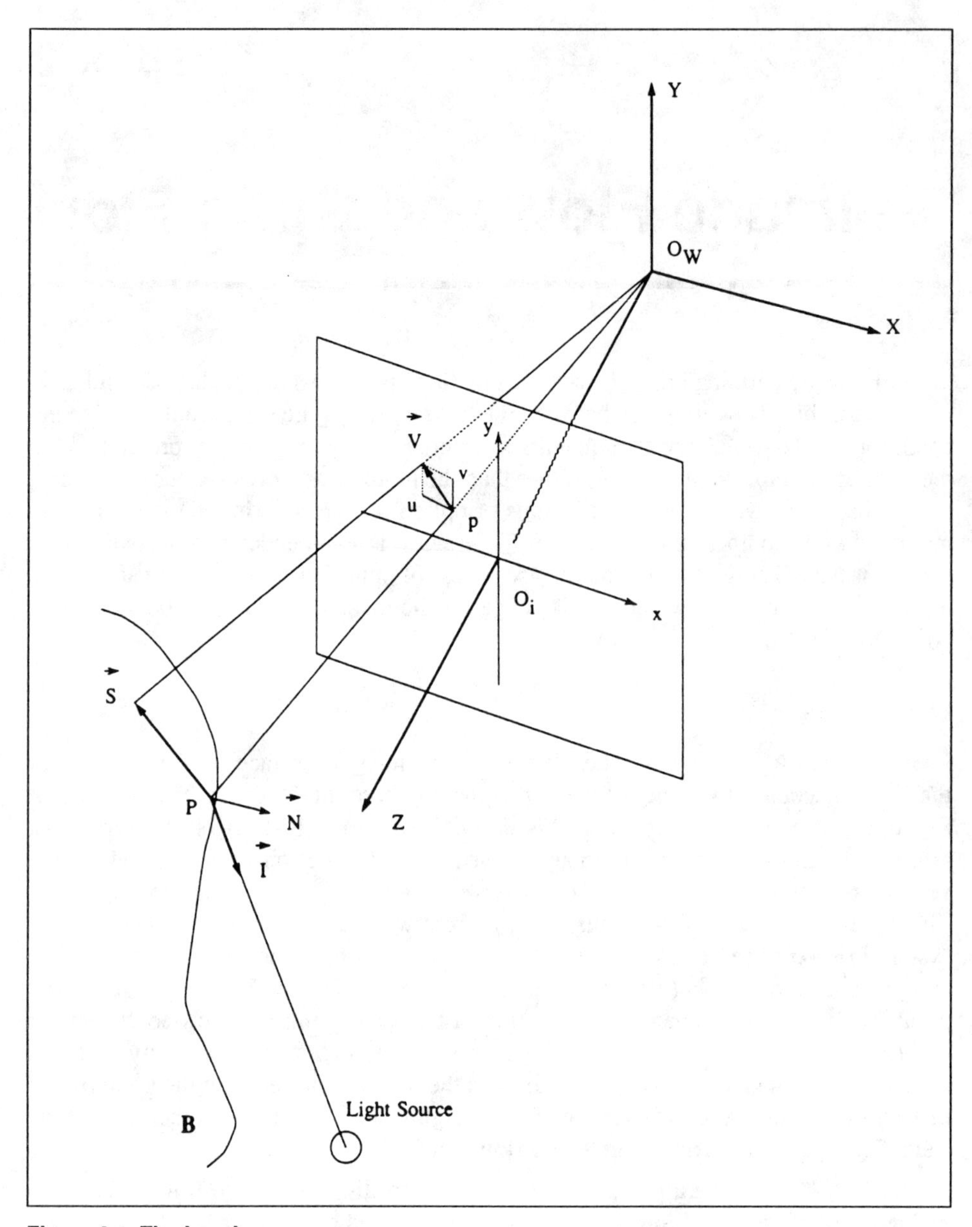

Figure 2.1. The imaging geometry.

could arise for several reasons; the simplest (to imagine) is a change in the scene illumination.

As the definitions imply, image flow and optic flow are generally not equal. For example, imagine a scene stationary with respect to the image plane and lit by a

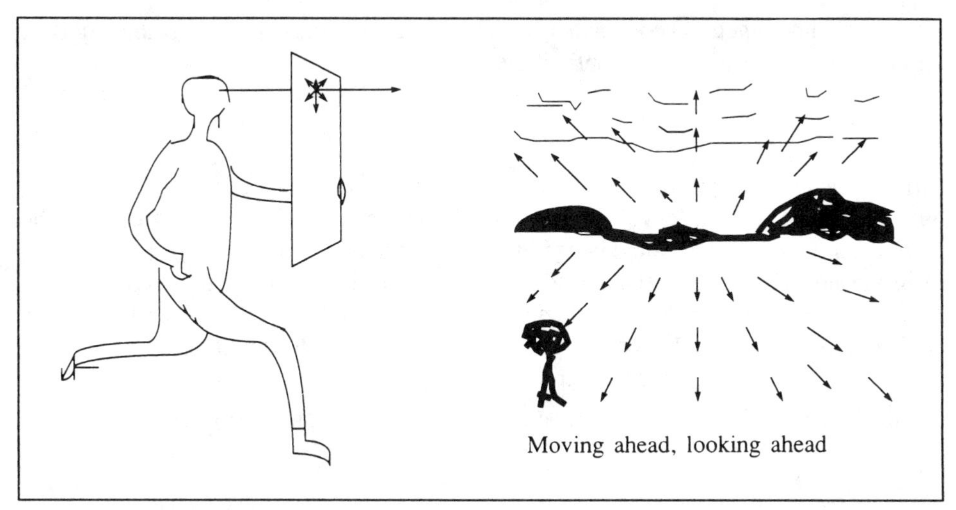

Figure 2.2 An example of image flow with a planar projection surface. (Redrawn from *Perception with an Eye for Motion*, 1986, by J.E. Cutting, by permission of The MIT Press, Cambridge, MA.)

nonuniform moving light source. Since the scene is stationary, the image flow will be zero throughout the image. However, since the light source is moving, the brightness patterns on the image will vary, resulting into a nonzero optic-flow field. The two quantities are equal only under a specific set of conditions.

The reason for deriving and understanding the conditions under which image flow and optic flow are equal is this: What we need for interpreting the 3D structure and motion of the scene is image flow, not optic flow. However, as the next chapter will reveal, all that is typically available is a sequence of intensity images of the scene, not the scene itself. The definitions above demonstrate that the best we can hope to recover—starting from the intensity images alone—is optic flow. We can then pretend that we have image flow and apply the image-flow interpretation algorithms to recover the 3D structure and motion. It is, therefore, important to find out exactly under what conditions the two quantities are equal.

The theorem of equality

> **Theorem 2.1.** The normal components of image flow and optic flow are equal if the time-varying imagery corresponds to a lambertian surface undergoing a purely translational motion under a spatiotemporally uniform illumination.

Verri and Poggio [106] have reported some general results for quantifying the difference between optic flow and image flow (they called it motion field) that can be applied to prove the theorem stated above. I will, however, give an alternative proof that is simpler and more intuitive. Before giving the actual proof, I review some definitions from basic photometry theory and give quantitative definitions of two more terms: normal image flow and normal optic flow.

A review of basic photometry. The *scene radiance* $\mathcal{L}$ is the power per unit area per unit solid angle emitted (or reflected) by a point in the scene, in a given direction. The *image irradiance* E (also called image intensity) is the power per unit area at a point (x, y) in the image. For Theorem 2.1, I deal only with *lambertian* or diffused surfaces. Hence, the terms scene radiance and image irradiance need to be defined for a lambertian surface. I use the imaging geometry of Figure 2.1 for these definitions and assume that the surface of rigid body B is lambertian.

The incident radiance arriving at the point $P = (X, Y, Z)$ on the surface is

$$\mathcal{L}_i \frac{watts}{m^2 ster^2}$$

The direction of the incident radiance is given by the unit vector **I**. The direction of the surface normal at the point P is given by the unit vector **N**. The (reflected) scene radiance $\mathcal{L}$ at this point is given by

$$\mathcal{L}(X, Y, Z) = \sigma \mathcal{L}_i \mathbf{I} \cdot \mathbf{N} \tag{2.1}$$

where σ is the constant of proportionality representing the surface albedo. Since the surface is assumed to be lambertian, the scene radiance at (X, Y, Z) is independent of the direction.

Further, for an image cast by a lambertian surface, the image irradiance $E(x, y)$ at the point $p = (x, y)$ is related to the scene radiance $\mathcal{L}(X, Y, Z)$ by a constant of proportionality κ as [107, 108]

$$E(x, y) = \kappa \mathcal{L}(X, Y, Z) \tag{2.2}$$

Normal image flow and normal optic flow. By definition, normal image flow is the component of the image flow along the local intensity-gradient direction. If $\mathbf{n}_E$ represents the unit vector in the direction of the local intensity gradient at a point in the image, the normal image flow v_n at that point is given by

$$v_n = \mathbf{n}_E \cdot \mathbf{V} \tag{2.3}$$

Using the vector notation,

$$\nabla \mathbf{E} = \left(\frac{\partial E}{\partial x}, \frac{\partial E}{\partial y} \right)$$

$$\mathbf{V} = (u, v) = \left(\frac{dx}{dt}, \frac{dy}{dt} \right) \tag{2.4}$$

we can write

$$v_n = \frac{1}{|\nabla E|} \nabla \mathbf{E} \cdot \mathbf{V} \tag{2.5}$$

which we can rewrite as

$$v_n = \frac{1}{|\nabla E|} \left(\frac{\partial E}{\partial x} \frac{dx}{dt} + \frac{\partial E}{\partial y} \frac{dy}{dt} \right) \tag{2.6}$$

Likewise, the component of *optic flow* along the local intensity-gradient direction is called normal optic flow and was shown by Horn and Schunck [34] to be

$$o_n = -\frac{1}{|\nabla E|} \frac{\partial E}{\partial t} \tag{2.7}$$

Subtracting Equation 2.7 from Equation 2.6, we can give the difference between normal optic flow and normal image flow as

$$v_n - o_n = \frac{1}{|\nabla E|} \left(\frac{\partial E}{\partial x} \frac{dx}{dt} + \frac{\partial E}{\partial y} \frac{dy}{dt} + \frac{\partial E}{\partial t} \right) \tag{2.8}$$

which we can rewrite as

$$v_n - o_n = \frac{1}{|\nabla E|} \frac{dE}{dt} \tag{2.9}$$

Proof of the theorem. The strategy to prove Theorem 2.1 is to express dE/dt in terms of more fundamental quantities and substitute it in Equation 2.9. Equations 2.1 and 2.2 can be combined to give

$$E = \kappa \sigma \mathcal{L}_i \mathbf{I} \cdot \mathbf{N} \tag{2.10}$$

Differentiating with respect to time, we can obtain

$$\frac{dE}{dt} = \kappa\sigma\left(\frac{d(\mathcal{L}_i\mathbf{I})}{dt}\cdot\mathbf{N} + \mathcal{L}_i\mathbf{I}\cdot\frac{d\mathbf{N}}{dt}\right) \tag{2.11}$$

The total differential of intensity with respect to time represents the difference between the intensities of two points on the image that are projections of a given point P on the moving surface at two different instants of time. As the surface moves, the albedo at point P on the surface remains constant. The orientations of the illumination unit vector **I** and the surface normal **N** with respect to the world coordinate system do change. Therefore, in computing the total derivative of intensity with respect to time, σ is not differentiated with respect to time, only $(\mathcal{L}_i\mathbf{I})$ and **N** are.

Substituting Equation 2.11 into Equation 2.9, we get

$$v_n - o_n = \frac{\kappa\sigma}{|\nabla E|}\left(\frac{d(\mathcal{L}_i\mathbf{I})}{dt}\cdot\mathbf{N} + \mathcal{L}_i\mathbf{I}\cdot\frac{d\mathbf{N}}{dt}\right) \tag{2.12}$$

Finally, from basic differential geometry [109], we obtain

$$\frac{d\mathbf{N}}{dt} = \Omega\times\mathbf{N} \tag{2.13}$$

where Ω is the rotational component of the scene velocity **S**. This, when substituted into Equation 2.12, gives

$$v_n - o_n = \frac{\kappa\sigma}{|\nabla E|}\left(\frac{d(\mathcal{L}_i\mathbf{I})}{dt}\cdot\mathbf{N} + \mathcal{L}_i\mathbf{I}\cdot\Omega\times\mathbf{N}\right) \tag{2.14}$$

Under the conditions of Theorem 2.1, illumination is assumed to be spatiotemporally uniform. That is, the total rate of change of illumination with respect to time is zero. With this assumption, we can rewrite Equation 2.14 as

$$v_n - o_n = \frac{\kappa\sigma\mathcal{L}_i}{|\nabla E|}\mathbf{I}\cdot\Omega\times\mathbf{N} \tag{2.15}$$

It is apparent that if Ω is zero, the difference between normal image flow and normal optic flow will be zero. That is, for a lambertian surface undergoing purely translational motion under a spatiotemporally uniform illumination, the normal components of image flow and optic flow are equal. This proves the theorem.

A corollary. From Equation 2.15, we can also show that

$$\lim_{|\nabla E| \to \infty} v_n - o_n = 0 \tag{2.16}$$

Thus, the normal components of image flow and optic flow are approximately equal in those regions of the image where the local intensity gradient has a very high magnitude.

As discussed earlier, the best we can hope to recover, given just the time-varying imagery, is optic flow, not image flow. The conditions under which the two are equal, as suggested by Theorem 2.1 and its corollary, can serve as a useful guideline in judging the reliability of a 3D interpretation process that uses an optic-flow field (which is an approximation of the image-flow field) as its input.

I prefer to use the term *estimation* rather than *measurement* (of optic flow). Measurement, by definition, refers to a direct "reading" by the sensor. According to this definition, what we measure is the image intensity, not optic flow. We *estimate* the optic flow from the measured intensity. The problem of computing optic flow becomes the classic problem of parameter estimation, the parameter being the optic-flow vector $\mathbf{V} = (u, v)$ at each point in the image. Using this model, we can draw useful conclusions about the confidence we can associate with the estimate. As later chapters will reveal, the idea of confidence measures is crucial.

Usefulness of optic flow

In this section, I review some potential applications of optic flow, broadly classified into 3D and 2D applications. The classic 3D application of optic flow is determination of the 3D structure and motion of the scene. Among the 2D applications, motion compensation has received much attention recently in the areas of video communication and medical imaging. Motion compensation is useful in enhancement and restoration of image sequences, and in image compression. Other 2D applications such as tracking, segmentation, and qualitative shape analysis, which I do not discuss here because of space limitations, are handled elsewhere [110].

Structure and motion from optic flow. From Figure 2.1 we can derive the geometric relationship between 3D structure and motion, and optic flow (assuming that it is equal to image flow). This relationship has been studied by a number of authors [39, 101, 111, 112]. I use a notation similar to that of Heeger and Jepson [112].

Assume that the position vector associated with the point P on the rigid body in Figure 2.1 is given by

$$\mathbf{P} = \begin{bmatrix} X \\ Y \\ Z \end{bmatrix} \tag{2.17}$$

and the velocity S of this point (that is, the velocity of the rigid body) is given by

$$\mathbf{S} = \begin{bmatrix} \frac{dX}{dt} \\ \frac{dY}{dt} \\ \frac{dZ}{dt} \end{bmatrix} = -(\Omega \times \mathbf{P} + \mathbf{T}) \tag{2.18}$$

where $\mathbf{T}$ and Ω are the translational and rotational components of the velocity, and are given by

$$\mathbf{T} = \begin{bmatrix} T_X \\ T_Y \\ T_Z \end{bmatrix} \qquad \Omega = \begin{bmatrix} \Omega_X \\ \Omega_Y \\ \Omega_Z \end{bmatrix} \tag{2.19}$$

Under perspective projection with focal length f, this point projects on to the point $p = (x, y)$ on the imaging surface, where

$$x = fX/Z \qquad y = fY/Z \tag{2.20}$$

The optic flow (truly speaking, the image flow), $\mathbf{V}$ ($= [u, v]^T$) can be obtained by differentiating Equation 2.20 with respect to time and substituting from Equation 2.18, and can be written as

$$\mathbf{V}(x, y) = p(x, y)A(x, y)\mathbf{T} + B(x, y)\Omega \tag{2.21}$$

where

$$p(x, y) = \frac{1}{Z}$$

$$A(x, y) = \begin{bmatrix} -f & 0 & x \\ 0 & -f & y \end{bmatrix} \tag{2.22}$$

$$B(x, y) = \begin{bmatrix} \frac{xy}{f} & -f - \frac{x^2}{f} & y \\ f + \frac{y^2}{f} & -\frac{xy}{f} & -x \end{bmatrix}$$

The matrices A and B depend only on the image position and the focal length. Assuming that optic flow is known, Equation 2.21 gives two bilinear equations in seven unknowns: the inverse depth $1/Z$, three components of the translational velocity $\mathbf{T}$, and three components of the rotational velocity Ω. (The image flow $\mathbf{V}$ is a linear function of $\mathbf{T}$ and Ω for a fixed p, and it is a linear function of p and Ω for a fixed $\mathbf{T}$.)

In principle, the 3D structure (the inverse depth for each image point) and motion (three components each of rotational and translational velocity of the rigid body with respect to the observer) can be determined if optic flow is available at five or more points in the image. However, a closer look at Equation 2.21 reveals the following. Since both $p(x, y)$ and $\mathbf{T}$ are unknown and are multiplied together in Equation 2.21, they can each be determined only up to a scale factor. In other words, we can determine only the direction of translation and relative depth (and the actual rotation).

Several researchers have attempted to use Equation 2.21 (or its variants) to compute 3D structure and/or motion under a variety of simplifying assumptions (such as known camera motion, purely translational but unknown camera motion, scene comprising only planar surfaces) [99, 102, 111, 112-121]. Aggarwal and Nandhakumar [103] and Barron [104] provide exhaustive and complete surveys of the literature on computing structure and motion from optic flow.

Motion compensation for image-sequence enhancement. Image sequences are used in a wide variety of applications: video communication, target detection and tracking, object recognition and localization, medical imaging, and so on. These image sequences are commonly corrupted by random noise at various stages—generation, transmission, or recording. The need to suppress this noise cannot be overstated. In applications such as video communication, the improvement of image quality offered by noise suppression is an advantage in itself. In target tracking and object recognition, this improvement in image quality enhances the robustness of low-level visual operators.

In medical applications such as X-ray fluoroscopy, image quality is closely linked to radiation dosage—the lower the dosage, the higher the noise. For a given radiation

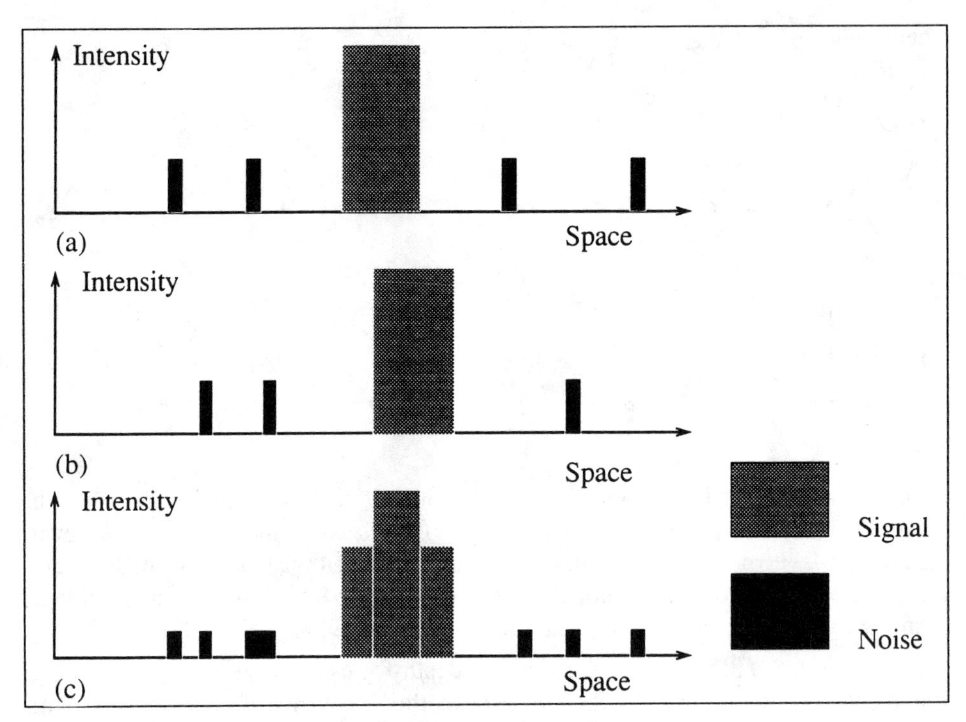

Figure 2.3. Temporal low-pass filtering blurs moving regions.

dosage, noise reduction can offer improved image quality, and hence, easier diagnosis. Also, for a given image quality, noise reduction allows the use of lower radiation dosage. This reduces the medical risks of exposure to radiation—for the patient and the physician alike.

In past research, temporal low-pass filtering has been commonly used to suppress noise in image sequences. Low-pass filtering reduces the statistical variance of noise. For instance, a mean filter that uses M samples (frames, in the case of temporal filtering) reduces the variance of additive white Gaussian noise by a factor of M. Similarly, a median filter, under similar conditions, reduces the noise variance by a factor of $2M/\pi$ [122]. It has been shown, however, that temporal low-pass filtering blurs the moving regions in the imagery. Figure 2.3 illustrates this for the case of a mean filter. Figures 2.3a and 2.3b show two frames of a one-dimensional image at two successive time instants. The averaged image in Figure 2.3c clearly shows blurring at the edges of the moving signal.

Motion compensation is commonly used to preserve moving regions during temporal smoothing [123, 124]. For this purpose, the optic-flow field is computed in advance for each frame involved in filtering. Then, all images except the current one are *warped* [125] using the flow field. If the flow field is accurate, the resulting images will have no interframe motion. These images are then fed to the low-pass temporal filter. Since there is no interframe motion, temporal filtering introduces no blurring. I will show a scheme for motion-compensated image-sequence enhancement in Chapter 7 and use it for noise suppression in some example sequences.

Chapter 3

State of the Art

Despite the numerous advances in neurophysiology, psychology, psychophysics, and computational vision over the last few decades, the problem of optic-flow estimation remains challenging. Its solutions are closer to an art than to a science. It remains a very active area of research in visual motion, and the state of the "art" is changing very rapidly. I will, nevertheless, attempt to review the various approaches that represent the current trends in optic-flow estimation. I am confident that many sections of this chapter will be obsolete by the time this book is published.

First, I give an explicit description of the problem of optic-flow estimation and show the physical basis of the well-known aperture problem that underlies it. Then, I identify three basic approaches that have been used to estimate optic flow. I briefly illustrate the underlying principle of each and show how the aperture problem appears in each formulation. Then, I review the past research in the aforementioned three approaches. Finally, I give a detailed analysis of the aperture problem and its possible solution.

Problem description

Figure 3.1 shows the problem of optic-flow estimation in a form most commonly encountered in computational vision. A finite sequence of images of a dynamic scene is available as an input to the *estimator*. The images are taken in quick succession so that the time elapsed between two consecutive images is small. The estimator must compute the optic-flow field corresponding to one of the input images. That is, it must assign a 2D instantaneous velocity vector to each pixel in a given input image. (In a real-time setting, the estimator accumulates a finite number of images up to the present time and must compute the optic-flow field corresponding to an image taken at some past instant.)

For example, an estimator using a sequence of three input images may compute the flow field corresponding to the central image. The restriction of *local computation* is usually imposed on the estimator. That is, while computing the velocity for a given pixel, the estimator must use information only in a small spatiotemporal neighborhood that surrounds it.

Many techniques reported in the past research use only two input images. They view the problem of optic-flow estimation as that of image-displacement estimation (with the assumption that the time elapsed between two images is unity). For each pixel in the first image, a "matching" or "corresponding" pixel must be found in the second image.

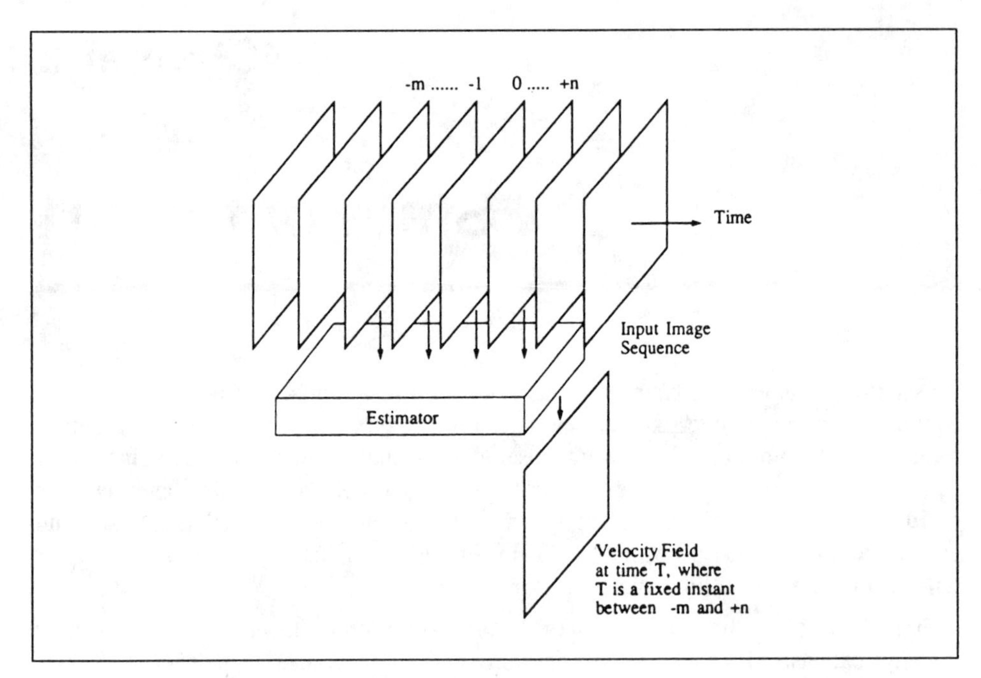

Figure 3.1. A description of optic-flow estimation.

I will use this definition to illustrate the aperture problem that underlies any technique for local estimation of optic flow.

The aperture problem. Figure 3.2 outlines the physical basis of the aperture problem. It shows a moving bar observed through an aperture that is small compared with the dimensions of the bar. Figure 3.2a shows the position of the bar at time t. Figure 3.2b shows the position of the bar at time $t + \Delta t$, assuming that the bar is moving eastward with a speed $\Delta x/\Delta t$. Figure 3.2c shows the position of the bar at time $t + \Delta t$, assuming that the bar is moving in the southeast direction with a speed $\sqrt{2} \times \Delta x/\Delta t$.

When we see the bar through the small the aperture alone, we cannot differentiate between the situation in Figure 3.2b and Figure 3.2c. More generally, it is not possible to compute the true velocity by observing a small neighborhood of a point that is undergoing motion. The only information *directly* available from local measurements is the component of the velocity normal to the underlying contour (commonly referred to as normal component or normal velocity). In the example, the normal velocity is $\Delta x/\Delta t$. The situation is quite different, however, if the aperture is located around one of the endpoints of the moving bar. Then we can directly compute the true velocity, because we know the exact location of the endpoint at two instants of time.

Generalizing, the aperture problem exists in regions of the image that have strongly oriented intensity gradients—for example, edges. On the other hand, image regions that have strong higher order intensity variations, such as intensity corners or textured

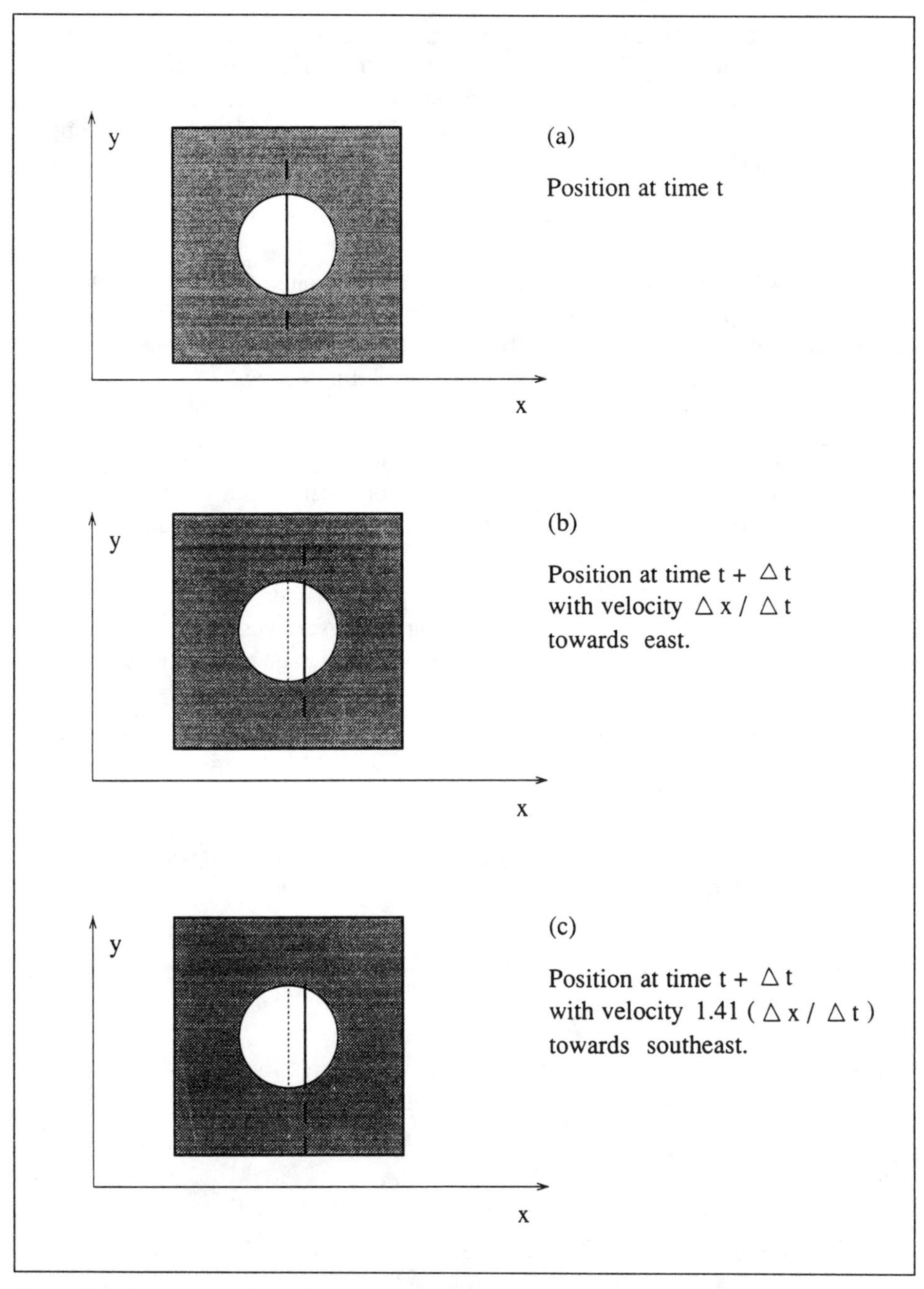

Figure 3.2. Physical basis of the aperture problem.

regions, do not suffer from the aperture problem. (This generalization, although sufficient for the current discussion, is rather crude. In Chapter 4, I replace it with a more elegant one.) The aperture problem is inherent in the problem of *local* estimation of optic flow and is not to be ascribed to any particular solution technique. This fact will reappear several times in this chapter.

Three approaches to optic-flow estimation

Most past work on optic-flow estimation has followed one of three approaches: the gradient-based approach, the correlation-based approach, or the spatiotemporal energy-based approach. In this section, I briefly explain the underlying principles of each and illustrate how the aperture problem appears in their formulations.

Gradient-based approach. Techniques based on the gradient-based approach [34, 67, 73, 74, 77] typically work on the assumption of *conservation of image intensity*. Some techniques use a function of image intensity rather than the image intensity itself. The underlying principle of estimation, however, remains the same.

These techniques assume that for a given scene point the intensity I at the corresponding image point remains constant over time. That is, if a scene point P (see Figure 2.1) projects onto the image point (x, y) at time t and onto the image point $(x + \delta x, y + \delta y)$ at time $(t + \delta t)$, we can write

$$I(x, y, t) = I(x + \delta x, y + \delta y, t + \delta t) \tag{3.1}$$

Expanding the right-hand side by a Taylor series about (x, y, t) and ignoring the second and higher order terms, we obtain

$$I(x + \delta x, y + \delta y, t + \delta t) = I(x, y, t) + \delta x \frac{\partial I}{\partial x} + \delta y \frac{\partial I}{\partial y} + \delta t \frac{\partial I}{\partial t} \tag{3.2}$$

Combining the two equations results in the following expression:

$$\delta x \frac{\partial I}{\partial x} + \delta y \frac{\partial I}{\partial y} + \delta t \frac{\partial I}{\partial t} = 0 \tag{3.3}$$

Dividing throughout by δt; denoting the partial derivatives of I by I_x, I_y, and I_t, and denoting the local velocity vector by (u, v), we get

$$I_x u + I_y v + I_t = 0 \tag{3.4}$$

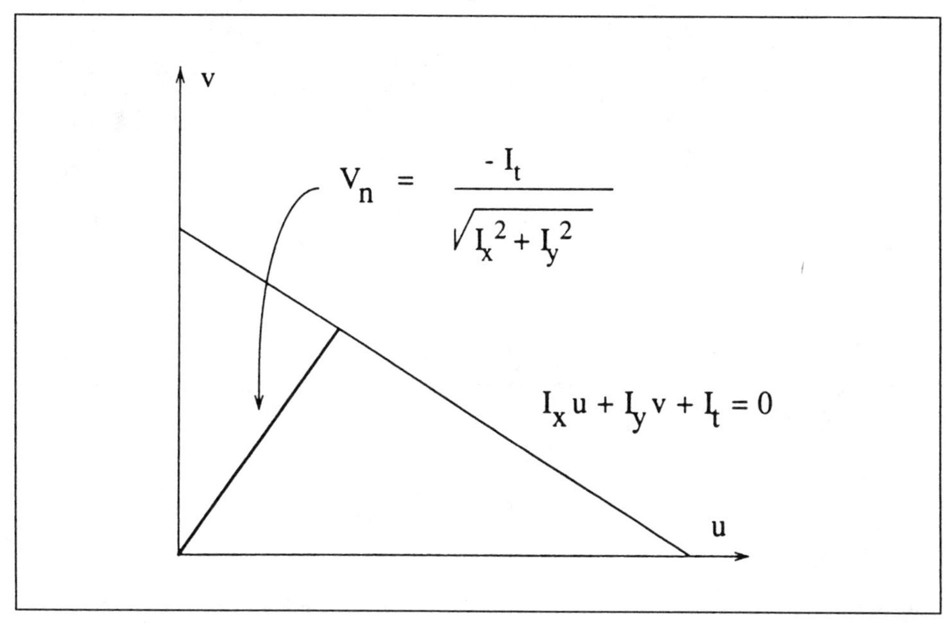

Figure 3.3. The motion constraint line.

Thus, the local constraints provide *one* linear equation in the variables u and v. As a consequence, the velocity vector (u, v) cannot be determined locally without applying additional constraints. Equation 3.4, also referred to as the *motion constraint line*, can be plotted in uv space, as shown in Figure 3.3. All that can be determined directly from Equation 3.4 is the normal velocity. This is equal to the perpendicular distance of the motion constraint line from the origin of uv space and is given by

$$V_n = \frac{-I_t}{\sqrt{(I_x^2 + I_y^2)}} \tag{3.5}$$

To summarize, the aperture problem appears in the gradient-based formulation as an underconstrained system composed of a single equation in two variables, namely, the two scalar components of the local velocity vector. This formulation is perfectly acceptable in the regions that do suffer from the aperture problem, that is, that have strongly oriented intensity gradients.

What about the regions that do not suffer from the aperture problem, such as corners? We should be able to recover the true velocity. It turns out that the simple gradient-based formulation described above cannot handle such regions. This is a serious drawback. The framework described above must be extended to extract the true velocity in the regions where the aperture problem does not exist and the normal velocity in the regions

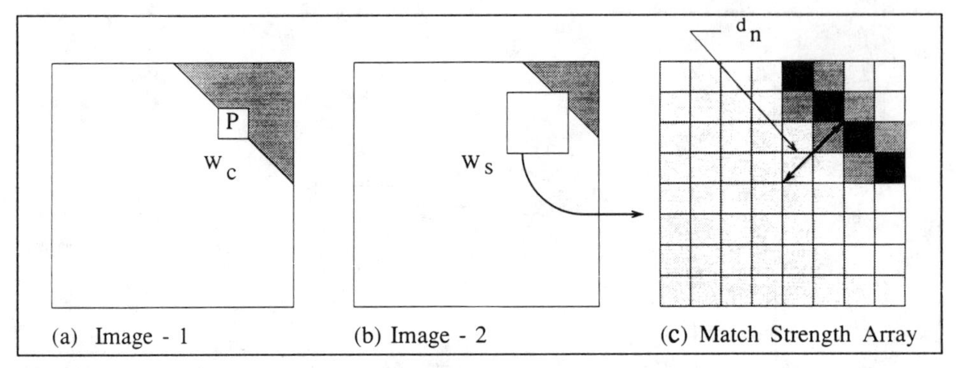

Figure 3.4. The aperture problem in correlation-based methods: (a) image 1, (b) image 2, (c) match strength array.

where it does. Later in this chapter, I will describe Nagel's technique, which satisfies this criterion.

Correlation-based approach. Correlation-based techniques typically use the assumption of *conservation of local intensity distribution.* They usually work on two successive images of the time-varying scene. In essence, for each pixel in the first image, they search for a matching pixel in the second image. The output is a displacement vector for each pixel in the first image. Typically, such a search involves finding the *best match* for the pixel under consideration (in the first image) among some *candidate* pixels in the second image. Most past research [52, 53, 81, 84, 126-128] uses the following strategies to determine the candidate pixels and the best match:

- The candidate pixels are determined by using a coarse-to-fine search strategy [80] or by imposing a physical constraint on the maximum displacement that can take place between two consecutive images in an image sequence. The region covered by the candidate pixels is called the *search window.*
- The best match is based on maximizing a *match measure.* The match measure is usually a correlation between a small *correlation window* around the pixel under consideration and a corresponding window around the candidate match. Using a small window (of a nonzero size) to compute the match measure follows from the assumption of conservation of intensity distribution in a local neighborhood of the pixel under consideration. Recently, Anandan [80] and Scott [83] have used weighted sum of squared differences between the intensities at the corresponding pixels in the two windows as a match measure. Finding the best match in this case amounts to minimizing the match measure.

If the best match can be found uniquely, the exact displacement vector is immediately known. This is true for points in the image that are sufficiently distinct, such as corners. Obviously, such points do not suffer from the aperture problem. The points in areas of strongly oriented intensity gradients such as edges, however, behave differently. This is illustrated in Figure 3.4. The correlation window W_c around pixel P in Figure 3.4a is matched against a similar window around each of the candidate pixels that lie in the search window W_s, shown in Figure 3.4b. The match measure is plotted over the search window, as shown in Figure 3.4c. Several points along the edge have values similar to the match measure. It is not possible to disambiguate among them. In other words, the aperture problem does exist.

In its most acute form (as in Figure 3.4), the aperture problem may lead to a scenario where several pixels in W_s having the same (highest) match measure show up along a straight line. The best that can be estimated in such a case is the component of the displacement vector along the underlying intensity gradient—that is, normal to the underlying intensity edge, depicted by d_n in Figure 3.4c. Additional constraints must be imposed to recover the true displacement vector.

Spatiotemporal energy-based approach. Much recent research has focused on the spatiotemporal frequency characteristics of a moving visual stimulus [85-88, 90-91] It is well understood [88] that the spatial and temporal frequencies are related to the velocity of a moving stimulus by

$$\omega_t = \omega_x u + \omega_y v \qquad \textbf{(3.6)}$$

Here, the variables u and v refer to the two orthogonal components of the 2D velocity vector, ω_x and ω_y refer to the two orthogonal components of the spatial frequency of the visual stimulus, and ω_t refers to the corresponding temporal frequency. According to Equation 3.6, if the spatial and temporal frequencies of a visual signal are plotted in $\omega_x\omega_y\omega_t$ space, they will lie on a plane. In other words, the energy of a moving stimulus is contained in spatiotemporal frequencies that lie on a plane in $\omega_x\omega_y\omega_t$ space. The image velocity (u, v) can be found by determining this plane.

Heeger [88] is the only one who addresses the problem of extraction of image velocity from the energy content of a moving visual stimulus. An underlying assumption of this work is that the visual stimulus is composed of imagery produced by a translating *textured* surface. Thus, it takes advantage of the fact that the aperture problem *does not* exist. In this case, the visual stimulus has many spatial frequency components (and the corresponding temporal frequency components) that can be observed. In terms of Equation 3.6, many points $(\omega_x, \omega_y, \omega_t)$ that lie on the motion plane are known. The plane is fully determined and both components of the velocity, u and v, can be found.

The situation is different, however, when the moving pattern is not textured. For example, a translating sinusoid may contain only a single set of spatial frequencies (ω_x^0, ω_y^0) which, when moving, give rise to a single temporal frequency ω_t^0. Thus we can characterize only two points, $(\omega_x^0, \omega_y^0, \omega_t^0)$ and $(0, 0, 0)$, in frequency space, and we cannot fully determine the motion plane. This situation represents the existence of the aperture problem. All that we can determine is a line that lies on the motion plane. In other words, the power spectrum is concentrated along a line in frequency space. The normal component of image velocity can be determined from the slope of this line. Again, we must impose additional constraints to determine the true velocity.

A generalization. It is apparent from the basic principles discussed above that the aperture problem appears in each of the three formulations of optic-flow estimation. This confirms the earlier assertion that the aperture problem is inherent in the problem of flow estimation. A technique to recover the true optic-flow field must overcome the aperture problem, explicitly or implicitly, by imposing additional constraints. Therefore, optic-flow estimation can be visualized as comprising two steps:

1. Set up some constraint(s) based on *conservation* of an image property, using one of the three basic approaches discussed above. Typically, this property is intensity [34, 68, 73, 74], some spatiotemporal derivative of intensity [39, 66, 129], or intensity distribution in a small spatial neighborhood [80-83]. Other choices are possible, for example, color. I refer to the first step as *deriving locally available velocity information in the form of conservation constraints.*
2. Use conservation constraint(s) along with some additional constraint(s) to recover true optic flow. This effectively solves the aperture problem. I show in a later section that the additional constraints are typically derived from the velocity distribution in a small neighborhood of the pixel under consideration. For this reason, I refer to the second step as *propagating velocity information using neighborhood constraints.*

The next two sections review the past research in optic-flow estimation in light of the two steps.

Conservation constraints: A review

Several recent survey articles [103, 104] review the past literature in detail. For each technique, they provide a complete description of the algorithm, the implementation details, and an evaluation of the overall performance. I review the past literature from a different perspective—to analyze how various techniques formulate the two steps described above.

In this section, I review some representative techniques and discuss how they approach the first step, namely, deriving the locally available velocity information in the form of conservation constraint(s). The review is divided into subsections devoted to the gradient-based approach, correlation-based approach, and spatiotemporal energy-based approach. In particular, I address the following three questions:

- Can the conservation constraint(s) recover the true optic flow in regions where the aperture problem does not exist?
- Can the conservation constraint(s) recover the normal optic flow in regions where the aperture problem does exist?
- Are there any inherent drawbacks in the conservation constraint(s) that make the implementation difficult or fragile?

Gradient-based approach. The techniques proposed by Horn and Schunck [34], Nagel [73, 74], Enkelmann [68], Buxton and Buxton [66], and Waxman, Wu, and Bergholm [78] illustrate the various ways conservation constraints can be derived using the gradient-based approach.

Horn and Schunck [34] assumed "conservation of intensity" and used Equation 3.4 as the conservation constraint. They further postulated that the rate of change of image intensity given by Equation 3.4 is not exactly zero because of quantization error and noise. The deviation from zero can be characterized by

$$E_b^2 = \left(\delta x \frac{\partial I}{\partial x} + \delta y \frac{\partial I}{\partial y} + \delta t \frac{\partial I}{\partial t} \right)^2 \tag{3.7}$$

The deviation E_b^2 must be minimized over the entire visual field.

Horn and Schunck also used the assumption of "smoothness of flow-field" to impose the additional constraint required to overcome the aperture problem. This corresponds to the second step and will be discussed in the section entitled "Neighborhood constraints: A review."

In terms of the three questions posed earlier, Horn and Schunck's formulation performs as follows:

- It takes no advantage of the fact that there are points in the visual field, such as corners, where the aperture problem does not exist and the true image velocity is completely known. The motion constraint provided by Equation 3.4 is incapable of capturing all the information available at a corner. In essence, the conservation constraint cannot extract the true optic flow in regions where the aperture problem does not exist.
- The formulation does give the normal optic flow everywhere in the visual field, according to Equation 3.5.
- In the actual experiments, Horn and Schunck used only two consecutive images of a synthetic image sequence and computed the partial derivatives using a $2 \times 2 \times 2$ spatiotemporal neighborhood. This implementation tends to be very fragile in real images. Furthermore, the technique is applicable only when the displacement over two consecutive images is small compared with the scale of spatial variations in intensity.

Nagel [73, 74] also assumed "conservation of intensity." He attributed the drawback of Horn and Schunck's formulation at the points of high curvature, such as corners, to the fact that the derivation of Equation 3.4 ignores the second and higher order partial derivatives of intensity. It assumes that the intensity varies linearly (with distance) in the local neighborhood of a point. In other words, it assumes that the local intensity surface at a point can be modeled as a plane. This assumption is particularly invalid in the vicinity of corners. Nagel attempted to overcome this anomaly by reformulating E_b^2 (in Equation 3.7) as

$$\iint E_b^2 dxdy = \iint (I(x,y,t) - I(x+u\delta t, y+v\delta t, t+\delta t))\, dxdy \tag{3.8}$$

and using up to second-order derivatives in modeling the intensity surface.

Nagel also introduced an additional constraint called "oriented smoothness" (that is minimized along with the conservation constraint shown above) to extract the true optic-flow field. This corresponds to the second step, and I discuss it in my review of neighborhood constraints later in this chapter.

Equation 3.8 alone can be used to derive a set of two linear equations in u and v given by the matrix equation [80]

$$[\mathcal{D}_s][V] = -[\mathcal{D}_t] \tag{3.9}$$

where the various matrices can be written in terms of the partial derivatives of intensity, the velocity (u, v), and a small spatial neighborhood σ, as follows:

$$[\mathcal{D}_s] = \begin{bmatrix} I_x^2 + I_{xx}^2 + I_{xy}^2\sigma^2 & I_x I_y + I_{xx} I_{xy}\sigma^2 + I_{yy} I_{xy}\sigma^2 \\ I_x I_y + I_{xx} I_{xy}\sigma^2 + I_{yy} I_{xy}\sigma^2 & I_y^2 + I_{xy}^2 + I_{yy}^2\sigma^2 \end{bmatrix}$$

$$[V] = \begin{bmatrix} u \\ v \end{bmatrix}$$

$$[\mathcal{D}_t] = \begin{bmatrix} I_x I_t + I_{xx} I_{xt}\sigma^2 + I_{xy} I_{yt}\sigma^2 \\ I_y I_t + I_{xy} I_{xt}\sigma^2 + I_{yy} I_{yt}\sigma^2 \end{bmatrix} \tag{3.10}$$

When the second-order derivatives are negligible, Nagel's scheme reduces to that of Horn and Schunck.

The conservation constraint used by Nagel addresses the three questions as follows:

- It can capture all the motion information available in the regions that do not suffer from the aperture problem (corners, etc.), giving the true optic flow directly [73].
- It can give normal optic flow in the regions that do suffer from the aperture problem, as in Horn and Schunck's method.
- It uses second-order partial derivatives of intensity that are difficult to compute in noisy real-life images. Although the scheme has no inherent restriction on the number of images used to compute the required temporal derivatives, Nagel shows

an implementation using only two successive images. Furthermore, this technique is applicable only when the amount of displacement over two consecutive images is small compared with the scale of spatial variations in intensity.

Enkelmann [68] used essentially the same conservation constraint as Nagel but used a hierarchical implementation. He described a pyramid data structure in which each image is a low-pass filtered version of the image at the next finer level, and he used a coarse-to-fine control strategy to transform the flow field computed at a coarse resolution to one at the next finer level. At each level, an iterative refinement of the flow field is performed using an oriented smoothness constraint similar to that of Nagel.

The coarse-to-fine strategy overcomes the drawback of Nagel's scheme with respect to the third question. The intensity derivatives computed at lower resolutions are more robust and give more reliable starting estimates of optic flow. Furthermore, small neighborhoods are quite appropriate for characterizing the intensity variations. Even though the restriction of motion must be small compared with the scale of intensity variation at each level of resolution, the actual motion that can be measured at the finest resolution does not have to be small.

Buxton and Buxton [66] deviated from the assumption of "conservation of intensity" used by Horn and Schunck, Nagel, and Enkelmann. They stated that conservation of intensity itself is an unrealistic assumption and postulated that conservation of some spatiotemporal derivative of intensity is more realistic. They chose the spatiotemporal d'Alembertian of Gaussian-smoothed intensity as the property to be conserved during motion. (Buxton and Buxton were not the first to use derivatives of the image intensity instead of the intensity itself as the conserved quantity. Marr and Ullman [40] used the Laplacian of the Gaussian of image intensity. However, their approach could find only the direction of normal velocity and not the magnitude.)

Buxton and Buxton first convolved the image sequence with a spatiotemporal Gaussian given by

$$G(x, y, t) = u\left(\frac{\alpha}{\pi}\right)^{\frac{3}{2}} e^{-\alpha(x^2 + y^2 + u^2 t^2)} \quad \textbf{(3.11)}$$

where the parameter α defines the spatial width of the Gaussian and the scale velocity u is introduced to maintain dimensional consistency. They computed the spatiotemporal d'Alembertian of the convolved output using the following operator:

$$\left(\frac{\partial}{\partial x^2} + \frac{\partial}{\partial y^2} - \frac{1}{u^2}\frac{\partial^2}{\partial t^2}\right) \quad \textbf{(3.12)}$$

Effectively, the conserved quantity $S(x, y, t)$ is given by

$$S(x, y, t) = m(x, y, t) * I(x, y, t) \qquad \textbf{(3.13)}$$

where $I(x, y, t)$ denotes the image sequence and the convolution operator $m(x, y, t)$ is given by

$$m(x, y, t) = \left(\nabla^2 - \frac{1}{u^2}\frac{\partial^2}{\partial t^2}\right) G(x, y, t) \qquad \textbf{(3.14)}$$

Buxton and Buxton postulated that $S(x, y, t)$ is conserved in the vicinity of contours corresponding to the intensity edges in the image (found as the zero crossings of $S(x, y, t)$). Thus, their scheme gives the normal velocity (at the edges) as

$$V_n = -\frac{\frac{\partial S}{\partial t}}{\sqrt{\left(\left(\frac{\partial S}{\partial x}\right)^2 + \left(\frac{\partial S}{\partial y}\right)^2\right)}} \qquad \textbf{(3.15)}$$

In their implementation, Buxton and Buxton chose the scale velocity u to be unity and used a $17 \times 17 \times 17$ spatiotemporal neighborhood to compute $S(x, y, t)$. They computed the partial derivatives of $S(x, y, t)$ using a $3 \times 3 \times 3$ neighborhood under the assumption that $S(x, y, t)$ varies linearly in the vicinity of a zero crossing.

The three questions relevant to this review have the following answers for Buxton and Buxton's formulation:

- No advantage is taken of the fact that there are points in the visual field where the aperture problem does not exist and the true image velocity is completely known. The constraint provided by conservation of $S(x, y, t)$ cannot capture the information available at a corner. In essence, the conservation constraint cannot extract the true optic flow in the regions where the aperture problem does not exist.
- The formulation does give the normal optic flow, but only at the intensity edges given by the zero crossings of $S(x, y, t)$.
- The implementation uses 17 frames of an image sequence and 17×17 spatial neighborhoods. Although expensive computationally, the scheme is very robust.

Waxman, Wu, and Bergholm [78] went a step further. They argued that although spatiotemporal intensity derivatives are a better choice than intensity itself, they are not truly conserved. Instead, they came up with what they called *activation profiles* whose shape, they claim, is truly conserved during motion. Their approach also computes normal velocity at the intensity edges, and is summarized (for a one-dimensional case) in Figure 3.5. Edge features in each input image are extracted as the zero crossings of the Marr and Hildreth edge detector, converting the gray-level image (Figure 3.5a) into

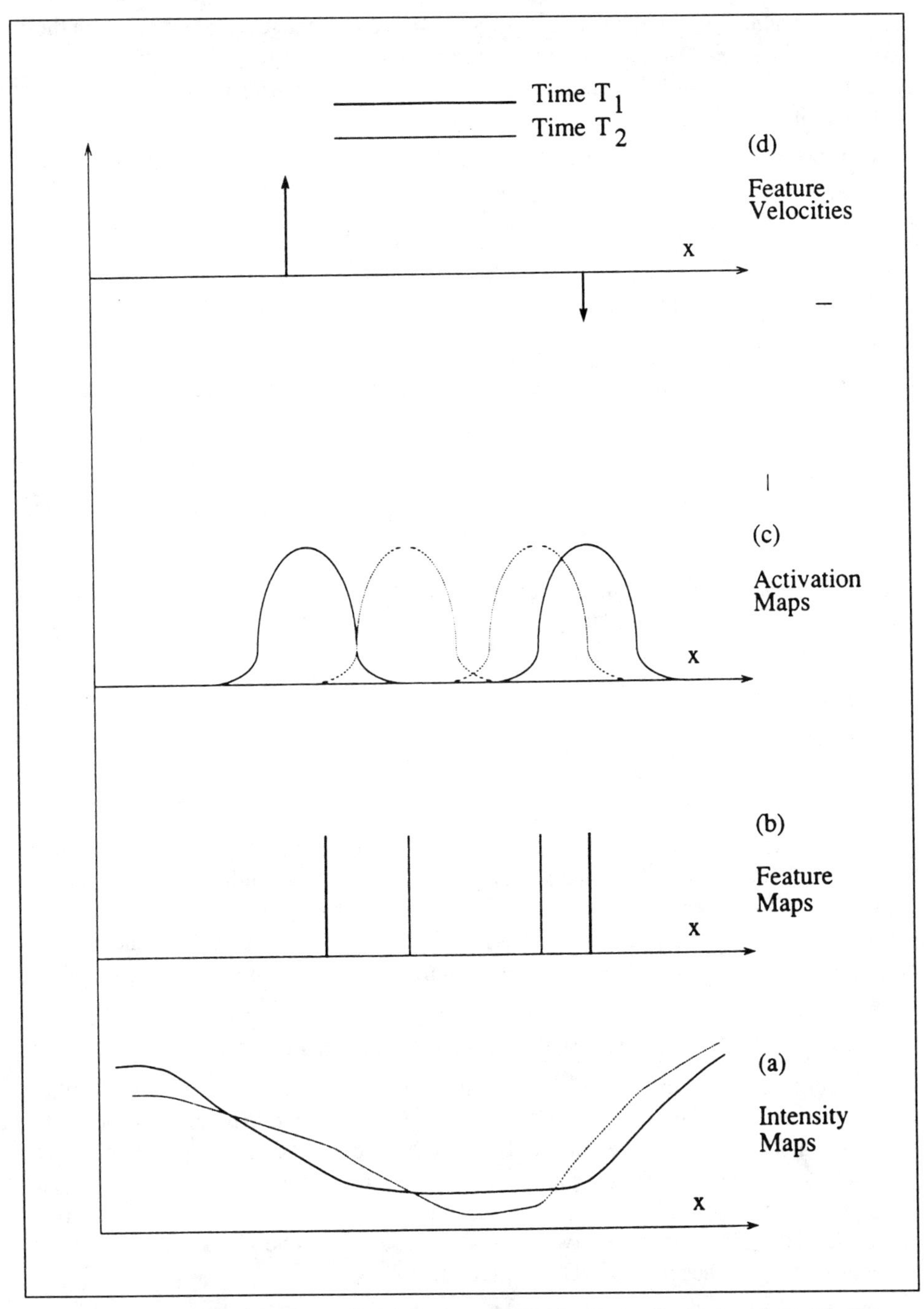

Figure 3.5. Velocity extraction from activation profiles in a one-dimensional signal. (Reproduced with permission from A.M. Waxman, J. Wu, and F. Bergholm, "Convected activation profiles and the measurement of visual-motion," *Proc. CVPR*, 1988.)

the corresponding binary image (Figure 3.5b). The binary images (feature maps, in their terminology) are convolved with spatiotemporal Gaussians to obtain *activation maps.* An activation map can be visualized as a gray-level image with a Gaussian intensity profile at the edge locations.

The processing required to convert the binary edge-image sequence $E(x, y, t)$ (feature-map sequence, in the original terminology) into the corresponding activation-gradient maps $A(x, y, t)$ is represented by

$$A(x, y, t) = G^{\sigma,\tau}(x, y, t) * E(x, y, t) \tag{3.16}$$

where the spatiotemporal Gaussian of amplitude C and spatial and temporal spreads σ and τ, respectively, are given by

$$G^{\sigma,\tau}(x, y, t) = Ce^{-\frac{x^2+y^2}{2\sigma^2}+\frac{t^2}{2\tau^2}} \tag{3.17}$$

With the assumption of conservation of activation $A(x, y, t)$, the normal velocity (at the edge locations) can be given by

$$V_n = -\frac{\frac{\partial A}{\partial t}}{\sqrt{\left(\left(\frac{\partial A}{\partial x}\right)^2 + \left(\frac{\partial A}{\partial y}\right)^2\right)}} \tag{3.18}$$

An implementation of this scheme was shown on an eight-stage PIPE machine [130-132]. The feature maps were found as zero crossings of the intensity image convolved with a difference of Gaussians (5 × 5 center, 11 × 11 surround). Activity maps were formed using a spatiotemporal Gaussian (9 × 9 pixels × 5 frames), and their partial derivatives were computed over the same neighborhood to obtain normal velocities at the edges.

With respect to the three criteria, this scheme performs as follows:

- The scheme of conservation of activation profiles at edges, by itself, cannot capture the true motion information available at corners. Waxman, Wu, and Bergholm did derive a scheme for conservation of activation profiles at corners that can extract true velocity. However, the scheme requires that the corners be extracted as points in the feature maps using a preprocessing stage and that the feature map contain *only* points. In essence, there are two separate frameworks, one for normal velocity at edges and the other for true velocity at corners.
- The formulation does give the normal optic flow, but only at the intensity edges.
- The assumption of conservation of the activity profiles is very reliable. However, there is an implicit assumption that the edges can always be detected and localized

Table 3.1. Deriving conservation constraints using the gradient-based approach.

Technique	Conserved Quantity	True Flow in Regions Free of Aperture Problem?	Normal Flow in Regions with Aperture Problem?	Other Comments
Horn and Schunck	Intensity	No	Yes	Uses only $2 \times 2 \times 2$ neighborhoods
Nagel	Intensity	Yes	Yes	Uses second-order derivatives
Enkelmann	Intensity	Yes	Yes	Hierarchical control
Buxton and Buxton	Spatiotemporal d'Alembertian of smoothed intensity	No	Yes	Very robust; works only at edges
Waxman, Wu, and Bergholm	Activation profiles	No	Yes	Very robust; works only at edges.

accurately and consistently, over various images. This is not true for a simple edge detector such as the one used in the original implementation.

Summary. Table 3.1 summarizes the performance of the five gradient-based schemes.

Correlation-based approach. Correlation has been used to establish feature correspondence in stereo images [133, 134]. In the case of stereo, corresponding features are spatially displaced in the two images, but one component of the displacement is known. In particular, the corresponding points in the two images lie along the *epipolar lines*. The use of correlation in optic-flow estimation was motivated by its similarity to stereo matching. The correlation-based approach can be used to find, for every point in one image, a matching point in the next image of the image sequence. This results in a displacement field. If the time elapsed between two images is very small, the displacement field is a good approximation of the velocity field (up to a scale factor).

In the section entitled "Three approaches to optic-flow estimation," I explained the underlying principle of deriving a conservation constraint using the correlation-based approach. It has been used to compute either sparse or dense displacement fields [53, 80-82, 84, 127, 128, 135]. In the following discussion, I examine three representative schemes—those of Wong and Hall [84], Burt, Yen, and Xu [81], and Anandan [80]—in light of the three questions posed in the beginning of this section. Because the formulation given by Anandan is similar to the framework that I develop in later chapters, I discuss it in detail.

Wong and Hall [84] used conservation of the local distribution of low-pass filtered intensity. They took two images as input and constructed a low-pass pyramid from each. At the lowest resolution, they formed a $(2n + 1) \times (2n + 1)$ correlation window around the pixel under consideration in the first image. They further assumed that the displacement of any pixel is not more than N pixels, giving a $(2N + 1) \times (2N + 1)$ search window in the second image.

There are several possible measures of the similarity between two patterns. Some commonly used measures are direct correlation, mean-normalized correlation, variance-normalized correlation, and sum of the squared differences [80]. Wong and Hall used *direct correlation* as the match measure between the correlation window and a similar subwindow around each pixel in the search window, giving a $(2N + 1) \times (2N + 1)$ array of matching strength. In their coarse-to-fine control strategy, they projected to the next finer level all the candidate pixels (in the search window) whose matching strength exceeded a certain threshold. The dimensions of both correlation window and search window were increased by a factor of two when moving from one level to the next finer level. At the finest resolution, the candidate pixel with the highest match measure was selected as the matching pixel.

Essentially, the scheme assumes that the intensity distribution around any pixel in a low-pass filtered version of the image is "carried along" as the pixel moves to a different location in the next image. The pixel $\mathcal{P}_2$ in the (low-pass filtered) second image, whose intensity-distribution is most similar to that of the pixel $\mathcal{P}_1$ under consideration in the first (low-pass filtered) image, is the best match for $\mathcal{P}_1$. Thus, the displacement vector for $\mathcal{P}_1$ is known.

The performance of this scheme with respect to the three basic questions is as follows:

- The scheme can find the true motion in the regions that do not suffer from the aperture problem.
- The conservation constraint is not strong enough to correctly recover the normal component of the motion in the regions that do suffer from the aperture problem. This is because the conservation constraint is based on maximizing the match measure. As explained in the discussion of the three approaches, such a maximum cannot always be found.
- The scheme is inherently based on two images.

Burt, Yen, and Xu [81] used conservation of the local distribution of band-pass filtered intensity. Using band-pass filtered images (instead of low-pass filtered images) makes the "matching-strength surface" (the distribution of the match measure over the search window) more distinct in the vicinity of features such as intensity edges and intensity corners [82]. Burt, Yen, and Xu constructed band-pass pyramids from two input images, used variance-normalized correlation as the match measure, and performed a hierarchical matching. The performance of the conservation constraint used in this scheme with respect to the three basic questions is exactly the same as that of Wong and Hall's scheme.

Burt, Yen, and Xu described a scheme (using confidence measures obtained from the directional second-order derivatives of the matching-strength surface) that works with the simple conservation constraint discussed above and recovers a good approximation to the true displacement field. Since this scheme is similar to a more general framework of computing and using confidence measures proposed by Anandan (described below), I will not discuss it here.

Anandan [80] also assumed conservation of the local distribution of band-pass filtered intensity and used a hierarchical framework. The processing involved at each level of the band-pass pyramid (he used Burt's Laplacian pyramid [136]) is summarized below.

For each pixel $\mathcal{P}(x, y)$ at the location (x, y) in the first (band-pass filtered) image, a correlation window $\mathcal{W}_p$ of size $(2 * n + 1) \times (2 * n + 1)$ is formed around the pixel. A search window $\mathcal{W}_s$ of size $(2 * N + 1) \times (2 * N + 1)$ is established around the pixel at location (x, y) in the second image. The (mis)match measure $\mathcal{M}(\delta x, \delta y)$ between $\mathcal{W}_p$ and a similar $(2 * n + 1) \times (2 * n + 1)$ window around each pixel in $\mathcal{W}_s$, displaced from (x, y) by an amount $(\delta x, \delta y)$, is computed as the sum of the squared differences as follows:

$$\mathcal{M}(\delta x, \delta y) = \sum_{i=-n}^{n} \sum_{j=-n}^{n} (I_1(x+i, y+j) - I_2(x+\delta x+i, y+\delta y+j))^2$$

$$-N \le \delta x, \delta y \le +N \qquad \textbf{(3.19)}$$

In this expression, $I_1(x, y)$ and $I_2(x, y)$ refer to the pixel intensities at the location (x, y) in the first and the second (band-pass filtered) images respectively. Also, the distribution of (mis)match measure over the search window is referred to as the SSD surface.

The pixel in $\mathcal{W}_s$ with the lowest (mis)match measure is selected as the best match, giving the displacement vector **D** for the pixel $\mathcal{P}(x, y)$. Further, the principal curvatures of the SSD surface at the best-match position are found, along with their respective directions. Anandan studied the behavior of the SSD surface and the principal curvatures in the vicinity of intensity corners, intensity edges, and homogeneous regions and concluded the following:

- In the vicinity of intensity corners, both principal curvatures are very large.
- In the vicinity of intensity edges, one principal curvature is very large and the other is very small. The principal direction corresponding to the smaller principal curvature has the same physical significance as the motion constraint line shown in Figure 3.3.
- In homogeneous regions, both principal curvatures are very small.

Anandan also suggested that the principal curvatures can be used as two directional confidence measures for the estimated displacement at a pixel and described a smoothing-based framework that uses the displacement vectors along with the directional confidence measures to estimate the displacement field at any level of the band-pass

Table 3.2. Deriving conservation constraints using the correlation-based approach.

Technique	Conserved Quantity	True Flow in Regions Free of Aperture Problem?	Normal Flow in Regions with Aperture Problem?	Other Comments
Wong and Hall	Local distribution of low-pass filtered intensity	Yes	No	Inherently based on two images
Burt, Yen, and Xu	Local distribution of band-pass filtered intensity	Yes	No	Hierarchical control
Anandan	Local distribution of band-pass filtered intensity	Yes	Yes	Gives confidence measures; uses hierarchical control

pyramid. He used a coarse-to-fine control strategy to propagate the displacement field to finer resolutions.

With respect to the three basic questions, Anandan's scheme performs as follows:

- The conservation constraint can directly recover the true displacement vector in the vicinity of corners, along with confidence measures in two orthogonal directions.
- In the regions that do suffer from the aperture problem, the normal component of the displacement vector can be recovered. It is equal to the perpendicular distance from the center of the search window $\mathcal{W}_s$ to the principal axis with the smaller principal curvature.
- Although the original implementation used only two images, extension to multiple frames is possible. Anandan alluded to a temporal coarse-to-fine strategy to accomplish this.

Summary. Table 3.2 summarizes the performance of the correlation-based schemes.

Spatiotemporal energy-based approach. As explained earlier, we can reduce the problem of optic-flow estimation to identifying a plane in spatiotemporal frequency space that has maximum energy associated with it. Much research has been done to explain the various theoretical and practical issues underlying the estimation of visual motion in the spatiotemporal frequency domain [85-88, 90, 91]. Only Heeger [88], however, went as far as developing a framework to estimate the optic flow. He implicitly used the assumption of "conservation of intensity" (Equation 3.6 is the frequency domain equivalent of Equation 3.4 in the space-time domain). Also, he assumed that the imagery is sufficiently textured, so the motion energy is distributed over several different spatiotemporal frequency channels. He used a set of Gabor filters to sample the motion energy in various channels and performed a least-squares estimation of the best-fit plane to compute the underlying velocity field. I summarize the overall scheme below.

A typical space-time Gabor filter is represented as the product of a space-time Gaussian and a sine (or cosine) wave:

$$g(x, y, t) = \frac{1}{\sqrt{2}\pi^{1.5}\sigma_x\sigma_y\sigma_t} \exp\left\{-\left[\frac{x^2}{2\sigma_x^2} + \frac{y^2}{2\sigma_y^2} + \frac{t^2}{2\sigma_t^2}\right]\right\} \sin(2\pi\omega_{x_0} + 2\pi\omega_{y_0} + 2\pi\omega_{t_0}) \quad (3.20)$$

where $(\omega_{x_0}\omega_{y_0}, \omega_{t_0})$ is the *center frequency*, that is, the spatial and temporal frequency for which the filter gives its maximum response, and $(\sigma_x, \sigma_y, \sigma_t)$ is the spread of the spatiotemporal Gaussian window.

The frequency response of a Gabor filter is given by

$$\begin{aligned} G(\omega_x, \omega_y, \omega_t) = {} & \frac{1}{4}\exp\{-4\pi^2[\sigma_x^2(\omega_x - \omega_{x_0})^2 + \sigma_y^2(\omega_y - \omega_{y_0})^2 + \sigma_t^2(\omega_t - \omega_{t_0})^2]\} \\ & + \frac{1}{4}\exp\{-4\pi^2[\sigma_x^2(\omega_x + \omega_{x_0})^2 + \sigma_y^2(\omega_y + \omega_{y_0})^2 + \sigma_t^2(\omega_t + \omega_{t_0})^2]\} \end{aligned} \quad (3.21)$$

implying that a filter with center frequency $(\omega_{x_0}, \omega_{y_0}, \omega_{t_0})$ will give a response equal to $G(\omega_x, \omega_y, \omega_t)$ when stimulated by a signal whose spatiotemporal frequency is given by $(\omega_x, \omega_y, \omega_t)$. Heeger used 12 Gabor filters tuned to center frequencies that lie along a cylinder in $\omega_x\omega_y\omega_t$ space, the axis of the cylinder being the ω_t axis. When stimulated by a 2D pattern translating with velocity (u, v), each of these filters will have a response that depends on the spatiotemporal frequency content of the pattern. A plane in $\omega_x\omega_y\omega_t$ space that gives the best fit to the outputs of the 12 filters defines the velocity (u, v) via Equation 3.6. Heeger used the technique of least squares to compute the best-fit plane.

Heeger's formulation behaves as follows with respect to the three basic questions:

- In absence of the aperture problem, the true velocity can be recovered directly.
- In the regions that do suffer from the aperture problem, the conservation constraint cannot directly recover the normal optic flow. Heeger does allude to constructing a response surface representing the distribution of energies over the 12 filters and computing the principal curvatures of the surface (around its maximum) to estimate the extent of the aperture problem and to compute the normal optic flow. He does not, however, provide a detailed analysis.
- Gabor filters are computationally very expensive. Several other alternatives have been suggested to construct spatiotemporally oriented filters [137], but their feasibility in optic-flow estimation is yet to be demonstrated.

Summary. I have discussed and evaluated various current techniques for deriving conservation constraints. On the basis of the three criteria used for evaluation, we can conclude that most of the current techniques do not satisfy the first two criteria simultaneously. That is, *either* the conservation constraint(s) cannot recover the true optic flow in aperture-problem-free regions *or* it (they) cannot recover the normal optic flow in the aperture-problem-affected regions.

The only exceptions are Anandan's and Scott's techniques, which use the correlation-based approach. They provide useful guidelines for the framework I develop in later chapters.

Neighborhood constraints: A review

The second step of the two-step solution that I proposed in my discussion of the three approaches to optic-flow estimation amounts to overcoming the aperture problem. The techniques currently used to overcome the aperture problem can be classified into two categories: They use either a smoothness constraint or the analytic structure of optic flow based on some assumptions about the underlying scene geometry. In this section, I review the representative techniques from both of these categories and evaluate their strengths and drawbacks.

This review does not provide a complete reference guide to the past work. Rather, it is a critical evaluation of some representative schemes. In particular, I address this question: How does the scheme perform at the discontinuities of optic flow? These discontinuities, typically referred to as the flow boundaries, arise in regions such as the object-background boundaries, depth discontinuities of moving rigid bodies, or occlusion boundaries of two or more objects moving at different velocities.

Smoothness constraint. Very commonly, the moving objects viewed in the world are opaque and undergo rigid motion or deformation. In this case, the neighboring points on the objects have similar velocities. Consequently, we can assume that the optic flow varies smoothly in small neighborhoods in the visual field. We can use this assumption to provide an additional constraint to the underdetermined system for optic-flow determination in the presence of the aperture problem. In the following paragraphs, I evaluate some formulations of the smoothness constraint.

Horn and Schunck [34] used the square of the magnitude of the gradient of optic flow E_c^2 as a measure of the *deviation* from smoothness in flow:

$$E_c^2 = \left(\frac{\partial u}{\partial x}\right)^2 + \left(\frac{\partial u}{\partial y}\right)^2 + \left(\frac{\partial v}{\partial x}\right)^2 + \left(\frac{\partial v}{\partial y}\right)^2 \qquad \textbf{(3.22)}$$

The additional constraint required to solve the aperture problem can be imposed by requiring that E_c^2 be minimized over the visual field (along with E_b^2 given by Equation

3.7). Thus, the problem of optic-flow estimation becomes that of satisfying two constraints at each point in the visual field. This can be achieved by minimizing:

$$E^2 = \int\int (\alpha^2 E_c^2 + (1-\alpha)^2 E_b^2)\, dxdy \tag{3.23}$$

In this function, α lies between 0 and 1 and can be interpreted as a smoothing factor. If α is very small, the resulting flow field will adhere very closely to the motion constraint. On the other hand, if it is very close to unity, it will tend to smooth out the flow field everywhere, including at genuine motion boundaries.

The principal objection to the formulation of Horn and Schunck is that it tends to blur the flow field at genuine motion boundaries. Essentially, the smoothness constraint has the effect of propagating the velocity information into the areas of uniform image intensity, where little or no information is available. This causes problems when an object moves against a uniform background, as it becomes difficult to distinguish between the velocities assigned to the object and the background.

Hildreth [71] proposed a technique to overcome the drawback of Horn and Schunck's formulation of the smoothness constraint. She extracted contours in the original intensity image using Marr and Hildreth's edge detector [138] and suggested that the velocity field be required to be smooth only *along* a contour and not across it. She proposed that the variation of the velocity **V** along a contour be measured as

$$\Theta(\mathbf{V}) = \int \left(\left(\frac{du}{ds} \right)^2 + \left(\frac{dv}{ds} \right)^2 \right) ds \tag{3.24}$$

where u and v denote the two scalar components of **V** and s denotes the arc length along the contour. Determining the velocity of a point along the contour is then formulated as an optimization problem to minimize $\Theta(\mathbf{V})$, subject to normal velocities measured at the contour points. Generally, the measurement of normal velocity is error prone. Hence, the function to be minimized is modified to include a term that denotes the difference between the computed and the measured values of the normal velocity at each point. The modified function is

$$\Theta(\mathbf{V}) = \int \left(\left(\frac{du}{ds} \right)^2 + \left(\frac{dv}{ds} \right)^2 \right) + \beta(\mathbf{V} \cdot \mathbf{n} - V_n))ds \tag{3.25}$$

Here, **n** is a unit vector normal to the contour at the point under consideration, V_n is the measured value of the normal velocity at that point, and β is a factor that represents the confidence in this measurement. The normal velocity measurements were done using a gradient-based formulation such as that of Horn and Schunck. The underlying approach, however, is general and is applicable irrespective of the technique used to compute normal flow.

Hildreth's formulation does not blur the flow field at motion discontinuities. But it suffers from the following drawbacks:

- It gives the true optic flow only at the points that lie along contours, not everywhere in the visual field.
- The underlying premise is that all the contours detected as edges in the image intensity correspond to motion boundaries. However, such contours can also be due to albedo transitions and texture. In such cases, velocity information at these contours should be propagated across the contour to provide an additional constraint on the velocities of intervening areas.

Nagel [73, 74] developed a technique that, at least in principle, overcomes most of the drawbacks of the smoothness-based formulations discussed so far. He observed that the smoothness error E_c used by Horn and Schunck is isotropic in the sense that it smooths out the flow field uniformly in all directions. The term corresponding to smoothness error in Horn and Schunck's original formulation (Equation 3.22) can be rewritten as

$$\int\int E_c^2 dxdy = \int\int \text{trace}\,((\nabla V)^T(\nabla V)\,dxdy \qquad \textbf{(3.26)}$$

The ∇V term expresses the partial derivatives of the two components of velocity in a matrix form

$$\nabla V = \begin{pmatrix} \frac{\partial u}{\partial x} & \frac{\partial v}{\partial x} \\ \frac{\partial u}{\partial y} & \frac{\partial v}{\partial y} \end{pmatrix} \qquad \textbf{(3.27)}$$

Nagel modified the smoothness error as

$$\int\int E_c^2 dxdy = \int\int \text{trace}\,((\nabla V)^T W(\nabla V)\,dxdy \qquad \textbf{(3.28)}$$

where W is a 2×2 positive-definite matrix defined as

$$W = \frac{F}{\text{trace}\,(F)} \qquad \textbf{(3.29)}$$

and the matrix F is given in terms of the partial derivatives of the intensity I and the local spatial neighborhood σ by

$$F = \begin{pmatrix} I_y^2 + \sigma^2(I_{xy}^2 + I_{yy}^2) & -I_x I_y - \sigma^2 I_{xy}(I_{xx} + I_{yy}) \\ -I_x I_y - \sigma^2 I_{xy}(I_{xx} + I_{yy}) & I_x^2 + \sigma^2(I_{xx}^2 + I_{xy}^2) \end{pmatrix} \qquad \textbf{(3.30)}$$

The change in smoothness error—that is, setting $W = F/\text{trace}(F)$ instead of an identity matrix used by Horn and Schunck—has the following effect. In regions with strong second-order intensity variations (for example, corners) the smoothness constraint is enforced very weakly: The flow field is allowed to be nonsmooth. Further, in the vicinity of edges (strong first-order but weak second-order intensity variations), the smoothness constraint is enforced strongly along the direction of the underlying contour and weakly across the contour. Because of these features, Nagel termed his smoothness constraint an *oriented* smoothness constraint. He computed the true velocity by minimizing:

$$E^2 = \int\int (\alpha^2 E_c^2 + (1 - \alpha)^2 E_b^2)\, dxdy \qquad \textbf{(3.31)}$$

where E_b and E_c are given by Equations 3.8 and 3.28 respectively.

Nagel's approach overcomes the drawbacks of both Horn and Schunck's and Hildreth's smoothness constraint. That is, it does not blur the flow field at motion boundaries and it gives the flow field everywhere, not just at the contours. However, it has practical limitations. It is based on the second-order spatial partial derivatives of the image intensity. Noise and quantization errors associated with practical imagery make the computation of second-order derivatives error prone.

Other techniques. Several techniques have been reported recently that incorporate an implicit measure to respect the motion boundaries during the smoothing process. Typically, this measure is based on an a priori confidence that two neighboring image points are the projections of two scene points that lie on a smooth surface. Notable examples are the schemes proposed by Kearney and Thompson [139], Aisbett [140], Hutchinson, Koch, and Mead [141], and Shulman [142].

Analytic structure of optic flow. If the geometry of the surface being imaged is known, projective transformations can be used to predict the nature of spatial variations in optic flow [44]. In this case, a heuristic such as the smoothness assumption is not needed to obtain the additional constraint required to overcome the aperture problem. Several researchers [98, 101, 102, 143] have used realistic assumptions about the underlying surface geometry to derive an analytic solution for optic flow. In this section, I review the approach taken by Waxman and Wohn [102], which is typical of such schemes.

Waxman and Wohn assume that there is a single moving rigid object whose velocity relative to the observer is

$$\mathbf{U} = -(\mathbf{V} + \Omega \times \mathbf{R}) \tag{3.32}$$

where $\mathbf{V}$ ($= V_X, V_Y, V_Z$) is the translational component of the velocity, Ω ($= \Omega_X, \Omega_Y, \Omega_Z$) is the rotational component of the velocity, and $\mathbf{R}$ ($= X, Y, Z$) is the position vector of the point on the object that corresponds to the point (x, y) in the image plane.

Under perspective projection, the image velocity (u, v) can be related to the image coordinates (x, y) by the following expressions:

$$u = \left(x\frac{V_Z}{Z} - \frac{V_X}{Z} \right) + (xy\,\Omega_X - (1 + x^2)\,\Omega_Y + y\,\Omega_Z)$$

$$v = \left(y\frac{V_Z}{Z} - \frac{V_Y}{Z} \right) + ((1 + y^2)\,\Omega_X - xy\,\Omega_Y + x\,\Omega_Z) \tag{3.33}$$

With the assumption that the viewed surface is locally planar, the reciprocal of Z is linearly related to the image coordinates by

$$Z^{-1} = Z_0^{-1}(1 - px - qy) \tag{3.34}$$

where Z_0 is the perpendicular distance to the plane from (0, 0) and $(p, q, -1)$ represents the direction of the normal to the plane. Substituting the value of Z^{-1} into the expressions for u and v, we obtain second-order polynomials in x and y. The constants in the polynomials can be interpreted as Taylor coefficients up to the second order (all higher order coefficients are actually zero):

$$u(x, y) = u(0, 0) + \frac{\partial u}{\partial x}x + \frac{\partial u}{\partial y}y + \frac{1}{2}\frac{\partial^2 u}{\partial x^2}x^2 + \frac{\partial^2 u}{\partial x \partial y}xy + \frac{1}{2}\frac{\partial^2 u}{\partial y^2}y^2$$

$$v(x, y) = v(0, 0) + \frac{\partial v}{\partial x}x + \frac{\partial v}{\partial y}y + \frac{1}{2}\frac{\partial^2 v}{\partial x^2}x^2 + \frac{\partial^2 v}{\partial x \partial y}xy + \frac{1}{2}\frac{\partial^2 v}{\partial y^2}y^2 \tag{3.35}$$

The normal flow at the point under consideration can then be given by

$$v_n(x, y) = n_x(x, y)u + n_y(x, y)v \tag{3.36}$$

Thus, at each point (x, y), we a have an equation in 12 unknowns, that is, the coefficients in a Taylor expansion. If we know the normal flow at 12 or more image points that correspond to the same planar patch on the imaged surface, we can determine the coefficients and compute the true velocity at all of these points using Equation 3.35.

Waxman and Wohn showed that the second-order model for optic flow is exact for a planar patch and gives a good local approximation for a quadratic patch. They tested their model on synthetic data and performed some perturbation tests to conclude that the model is quite robust.

Regarding the performance of this scheme at motion boundaries, the normal flow measurements used to compute the 12 coefficients must correspond to a local patch lying on a single surface. More specifically, the local neighborhood used in the computations must not span a surface discontinuity. This requires a reasonable high-level image segmentation (beyond a simple zero-crossing analysis). Waxman and Wohn suggested that the second-order model itself can be used to find the boundaries of analyticity (boundaries across which the second-order model ceases to hold true). They showed that this scheme works very well for synthetic imagery. However, its performance with real imagery remains to be established.

Summary. The techniques reviewed in this section use velocity information in a small neighborhood of the point under consideration to derive the additional constraint(s) required to overcome the aperture problem. For this reason, I have referred to the additional constraint(s) as neighborhood constraint(s). From the discussion of various techniques to obtain the neighborhood constraint(s), we can conclude that they suffer from one or more of three drawbacks:

- they tend to blur the flow field across the motion boundaries, or
- they use the second-order derivatives of optic flow, which are hard to estimate from noisy image data, or
- they assume that image segmentation has been done, thus requiring a significant amount of global information.

Conclusion

This chapter subdivided the problem of optic-flow estimation into two logically distinct steps and reviewed the formulation of these two steps in various algorithms reported in past research. The reason for this new analysis is that most of the previously published review articles [103, 104] examine the various algorithms "on the whole." For a given algorithm, some inherent drawback in the formulation of the first step may be "covered up" by the second step, making the drawback invisible or only partly visible at the output stage. The review in this chapter can help unravel such hidden drawbacks and identify a strong formulation for each of the two steps—to design a stronger overall algorithm. I attempt to design one such algorithm in the following chapters.

Chapter 4

Estimation-Theoretic Framework: Preliminaries

In this chapter, I introduce some preliminaries of a new framework for recovering optic flow from time-varying imagery. I call this framework estimation-theoretic because it views the problem of optic-flow recovery as a parameter-estimation problem and applies some of the techniques commonly used in estimation theory to compute optic flow.

In Chapter 3, I mentioned that the process of optic-flow recovery underlies two functional steps. The objective of the first step is to recover whatever velocity information is available locally, everywhere in the visual field. (*Locally available information* is the information that can be recovered by using very small spatial and temporal neighborhoods.) This information is recovered in the form of conservation constraints based on the assumption of conservation of some image property over time. Furthermore, this information is not complete everywhere in the visual field. In several regions, there may only be partial information or no information at all. The objective of the second step is to propagate velocity from regions of full information to regions of partial or no information, in order to recover the correct flow field. Velocity propagation is achieved by using neighborhood constraints. Summarizing, in the framework shown in Chapter 3, the information used to recover optic flow is in the form of conservation constraints and neighborhood constraints, both obtained from a small spatial and temporal neighborhood around the pixel under consideration.

The objective of this chapter is to introduce a more general notion of information that can be derived from small spatiotemporal neighborhoods in time-varying imagery, for the purpose of recovering optic flow. First, I discuss the nature of information that can be recovered by assuming conservation of some image property over time. I call it *conservation information.* I show that conservation information is not exact—there is always uncertainty associated with it. I discuss two equivalent representations for this information:

- An *estimate* along with a *covariance matrix.*
- Two linear constraints, each with an associated confidence measure. (I discuss the assumptions under which these representations are valid in Chapter 5.)

Then, I discuss the information used in the second step—velocity propagation. This information, referred to as *neighborhood information*, is derived from the knowledge of velocities of points in a small neighborhood around the pixel under consideration. I show that this information can be represented exactly like conservation information.

Conservation information

According to the definition given earlier, conservation information is the information (about optic flow) that can be recovered by local measurements alone, by assuming conservation of some image property over time. Obviously, this information can at best be equal to the information that is actually available in small spatiotemporal neighborhoods. In the following discussion, I revisit the aperture problem to explain the issue of local availability of velocity information. My objective is to ascertain exactly how much information about optic flow is available in a small spatiotemporal neighborhood.

Traditionally, the aperture problem has been regarded as a "binary" problem—either it exists or it doesn't. This view of the aperture problem was presented in Chapter 3 using the "translating edge" experiment. However, because imagery is inherently noisy, it is more illustrative to treat the aperture problem as a continuous one. It always exists, but in varying degrees of acuteness. I demonstrate this using the following example. Imagine each of the six patterns shown in Figure 4.1 moving and being observed through a small aperture. We can make the following observations about the velocity information that can be recovered when the patterns are viewed through the small aperture alone:

- Pattern 1 shows a moving high-contrast corner, free of any noise. The position of the corner point can be found exactly at various time instants during which it is visible through the aperture. Thus, the exact velocity of the pattern is known. We can plot this velocity in uv space as a single point. (Here, u and v are the two orthogonal components of the image velocity).
- Pattern 2 shows a similar moving corner, but with noise introduced by digitization or some other factor. The acuity is lower in all directions and the location of the corner point is relatively "fuzzy" at various time instants. Therefore, the velocity of the pattern is not known exactly, but we can assume that it lies in a small region of uncertainty in uv space. (The uncertainty in locating the corner results from noise. If the pattern represents a noise-free blurred corner, the correct velocity can still be recovered. The figures show the regions of uncertainty as circular or elliptical only for simplicity. In reality, they can have more complex shapes.)
- Pattern 3 shows a similar corner in which the effects of noise are unequal in the horizontal and vertical directions. The horizontal acuity is the same as that in Pattern 2, but vertical acuity has been reduced further. The resulting region of uncertainty in which the velocity lies in uv space will have a larger extent in the vertical direction than in the horizontal direction.
- Pattern 4 is derived from Pattern 3 by completely eliminating the vertical acuity, giving a fuzzy edge. Following the argument given for Pattern 3, we can say that the region of uncertainty in uv space has an infinite extent in the vertical direction.

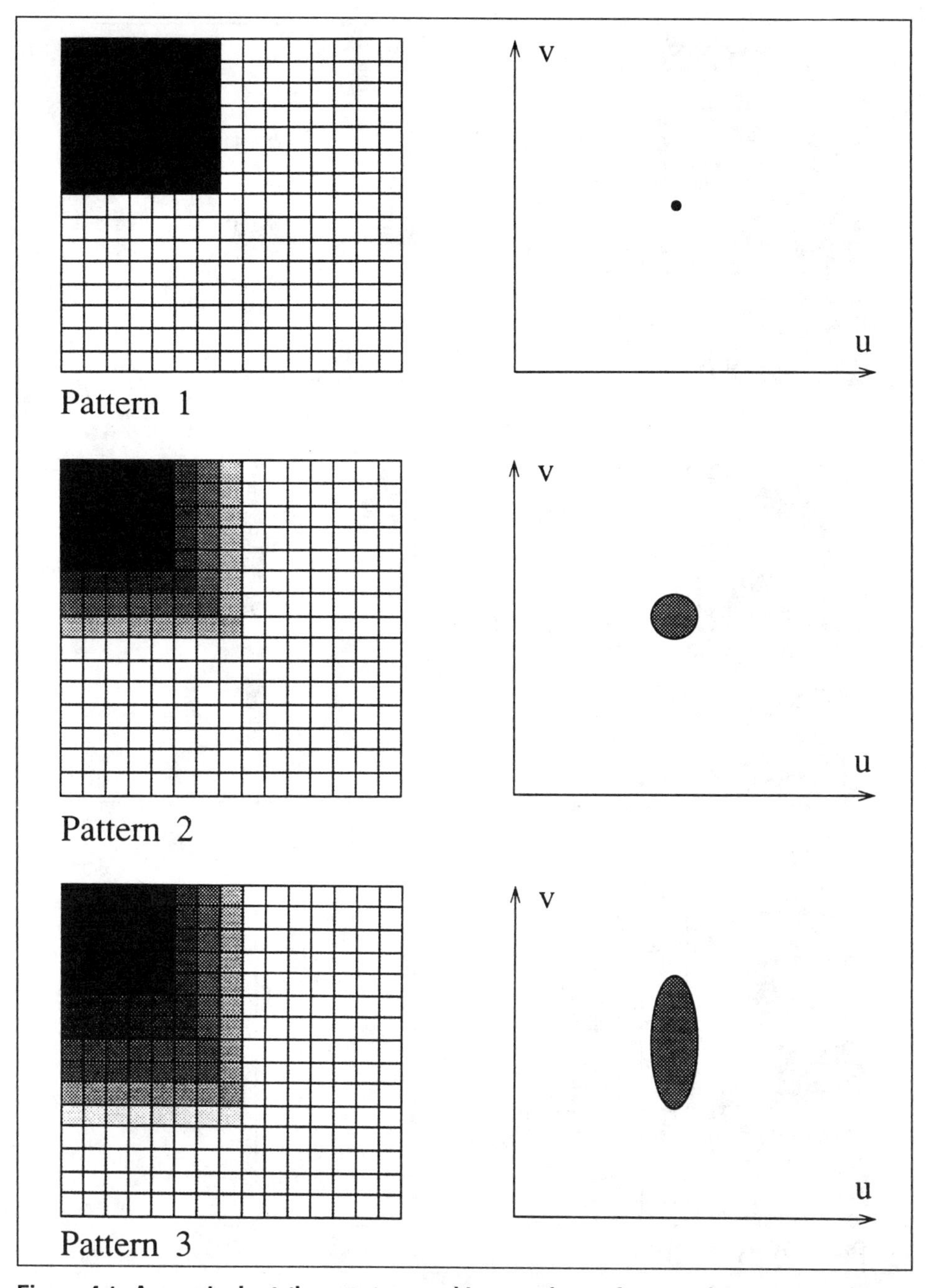

Figure 4.1. A new look at the aperture problem: regions of uncertainty associated with velocity information when various patterns are viewed through a small aperture (figure continued on the next page).

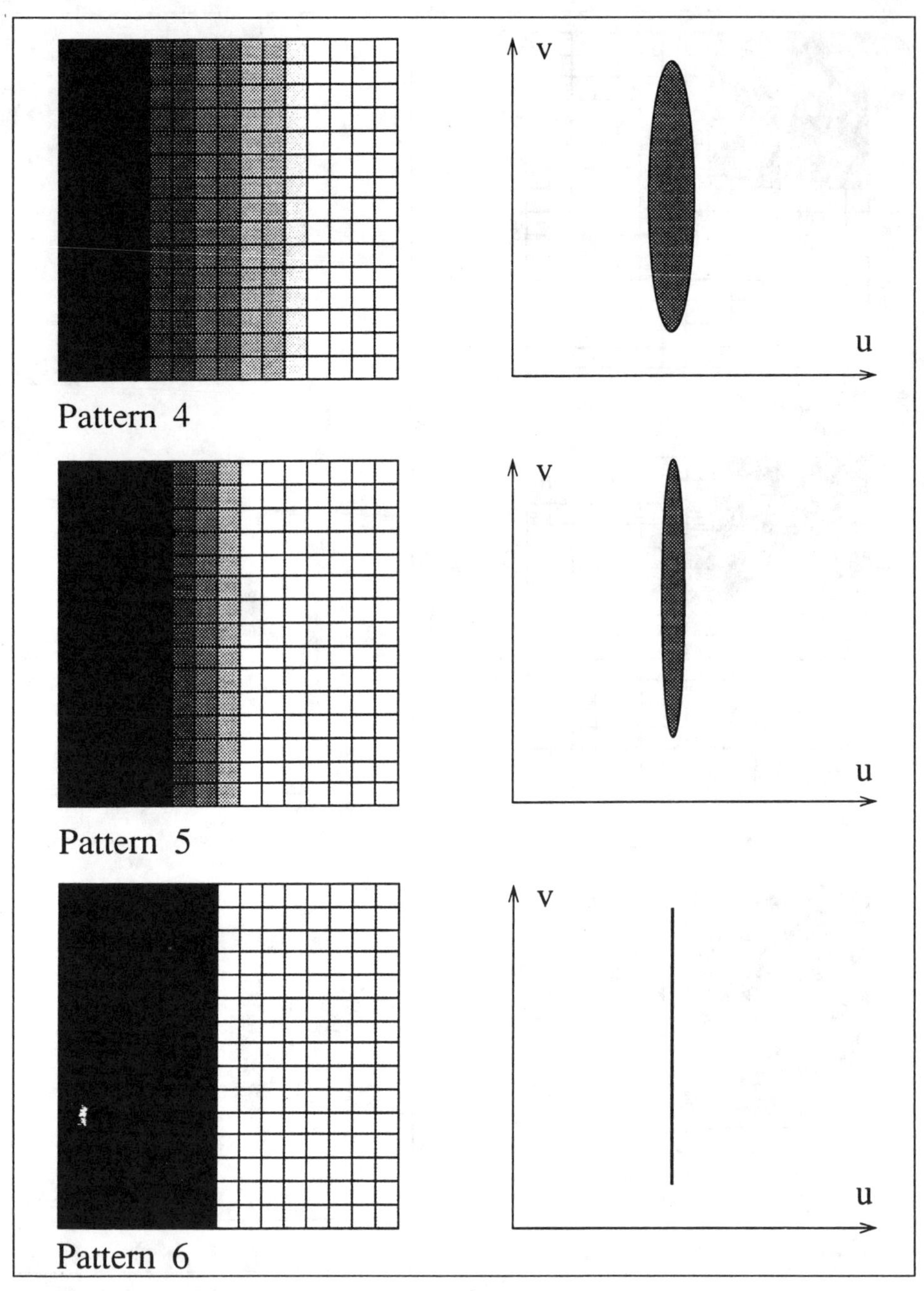

Figure 4.1, continued. In patterns 4, 5, and 6, the actual extent of the region of uncertainty is infinite in the vertical direction.

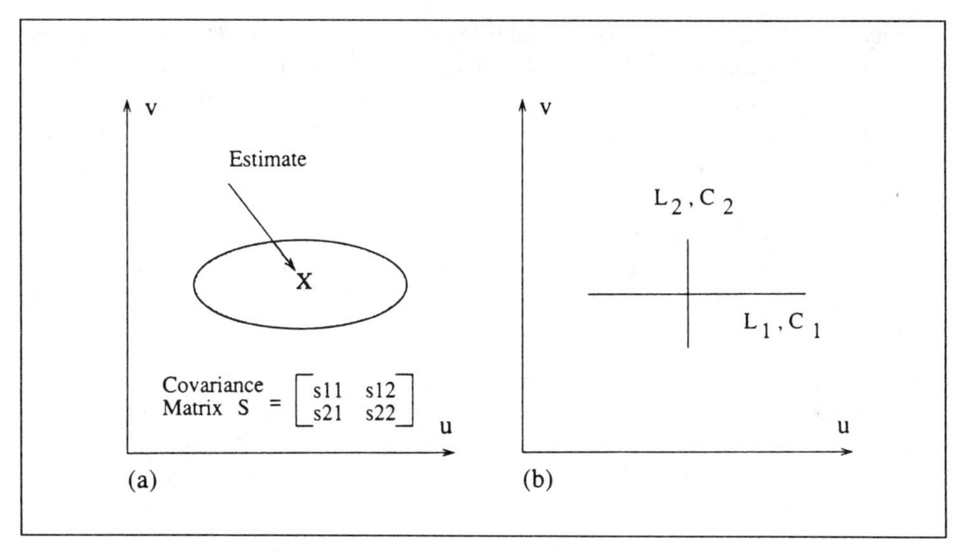

Figure 4.2. Two representations of the locally available velocity information: (a) an estimate and a covariance matrix, (b) two linear constraints with their associated confidence measures.

- Pattern 5 is obtained from pattern 4 by decreasing the effects of noise in the horizontal direction. This makes the horizontal extent of the region of uncertainty in *uv* space narrower.
- Pattern 6 is obtained from pattern 5 by eliminating noise completely, giving a sharp edge. This makes the horizontal extent of the region of uncertainty in *uv* space infinitesimally small, giving a straight line along which the velocity must lie. This is consistent with the traditional view of the aperture problem.

Generalizing the observations of this experiment, we can say that the information about the image velocity that is *available* in a spatiotemporally local neighborhood (i.e., that can be *recovered* by spatiotemporally local measurements alone) is never complete. Given some assumptions, we expect the velocity at a point in the image to lie in a region of uncertainty in *uv* space (see Figure 4.2a).

The extent of the region of uncertainty depends on the underlying intensity variations in the imagery. There are several ways to quantify this kind of information. A representation commonly used in estimation theory is an estimate accompanied by a covariance matrix [144]. I will show in Chapter 5 that, under some realistic assumptions, velocity information recovered by local measurements can indeed be represented in this form.

Alternatively, we can represent this information as follows (see Figure 4.2b). Let $\mathcal{L}_1$ and $\mathcal{L}_2$ denote two straight lines aligned with the principal axes of the region of uncertainty for the velocity at a given point in the image. We can think of these lines as (linear) constraints on velocity. Furthermore, we can think of the reciprocal of the "spread" about each of these lines as a measure that describes the confidence in the true velocity lying along that line—the smaller the spread, the higher the confidence. I denote the confidence associated with the two constraints C_1 and C, respectively. This representation was used by Scott [135], and it ties very well with the generalization proposed in Chapter 3. The two linear constraints can be thought of as *conservation constraints.* I use the first representation in most of the book, occasionally showing the connections with the second representation.

The preceding discussion pertains to the velocity information that is available in a small spatiotemporal neighborhood. This information has to be recovered from the image sequence by assuming conservation of some image property over time. The desirable features of a recovery procedure are as follows:

- It should be possible to recover *all* the local information available in any one of the two representations, using any one of the three basic approaches: the gradient-based approach, the correlation-based approach, and the spatiotemporal energy-based approach. Furthermore, the assumptions and the procedures for recovery of this information must be similar for all three approaches.
- The recovered information should correctly reflect the available information. For example, in the vicinity of an intensity corner or in an isotropically textured region, both conservation constraints should have a high confidence. On the other hand, in the vicinity of an intensity edge, in a continuous-intensity region that has a strongly oriented intensity gradient, or in an anisotropically textured region, one constraint should have a high confidence while the other should have a low confidence. Likewise, in a dull region, both the constraints should have a low confidence.

Most of the techniques reviewed in the previous chapter do not satisfy the above criteria. For example, the classic Horn and Schunck procedure [34] gives only one of the two available conservation constraints. (Horn and Schunck refer to this constraint as the motion-constraint line. It follows from the discussion of pattern 6 in the experiment shown in Figure 4.1 that this constraint has the same physical meaning as the major axis of the region of uncertainty—that is, as the high-confidence constraint.) Horn and Schunck's procedure does not give any confidence measure associated with the conservation constraint.

Similarly, the techniques of Buxton and Buxton [66] and Waxman, Wu, and Bergholm [78] give only one conservation constraint, without any confidence measure, and only at the edge locations. Nagel's method [73, 74] does give a pair of linear constraints on velocity at each point in the image. However, it does not have any explicit notion of confidence measures. The techniques of Anandan [80] and Scott [83] that use the correlation-based approach are among the very few that recover local velocity information along with explicit confidence measures.

In Chapter 5, I will describe a procedure to recover local velocity information in each of the two representations described earlier using the correlation-based approach. In Chapter 8, I will show that a similar procedure can be applied to recover local information in these representations using either of the remaining two approaches: the gradient-based approach and spatiotemporal energy-based approach.

Neighborhood information

Before I describe the nature of neighborhood information, I will briefly discuss the need for velocity propagation and the desirable features of a propagation procedure. The preceding section shows that locally available information about image velocity is not complete everywhere in the visual field. In many regions, the information is only partial. In terms of conservation constraints, there may be regions in the image where both constraints have a high confidence, regions where only one constraint has a high confidence, and regions where both constraints have a low confidence. Velocity must be propagated from regions of high confidence to regions of low confidence. The discussion in Chapter 3 demonstrated that the propagation procedure must use conservation information at the pixel under consideration as well as neighborhood information. The features desirable in a velocity-propagation procedure are the following:

- In making use of conservation information at the pixel under consideration, it must explicitly use confidence measures. This is important to ensure that the estimates with a high confidence stay unaltered during the propagation procedure and those with a low confidence become more accurate and acquire high confidence as propagation progresses.
- In making use of neighborhood information, the procedure should minimally blur the flow field at genuine motion boundaries. Also, it should not require a priori knowledge about the location of motion boundaries. Instead, it should generate boundaries as a side effect.

To illustrate the nature of neighborhood information, I review a classic way in which it has been used—Horn and Schunck's smoothing-based propagation procedure [34]. (I discussed their original formulation in Chapter 3.) Horn and Schunck used a calculus of variations to determine the conditions that minimize the error given by Equation 3.23 and derived the following iterative solution, with (u^n, v^n) denoting the velocity of a given pixel at nth iteration:

$$
\begin{aligned}
u^{n+1} &= \overline{u^n} - I_x \frac{I_x \overline{u^n} + I_y \overline{v^n} + I_t}{\alpha^2 + I_x^2 + I_y^2} \\
v^{n+1} &= \overline{v^n} - I_y \frac{I_x \overline{u^n} + I_y \overline{v^n} + I_t}{\alpha^2 + I_x^2 + I_y^2}
\end{aligned}
\qquad \textbf{(4.1)}
$$

After a slight rearrangement and after removing the superscripts—that is, after considering a steady-state scenario—we can rewrite the solution as

$$
\begin{aligned}
(\alpha^2 + I_x^2 + I_y^2)(u - \bar{u}) &= -I_x(I_x\bar{u} + I_y\bar{v} + I_t) \\
(\alpha^2 + I_x^2 + I_y^2)(v - \bar{v}) &= -I_y(I_x\bar{u} + I_y\bar{v} + I_t)
\end{aligned}
\qquad \textbf{(4.2)}
$$

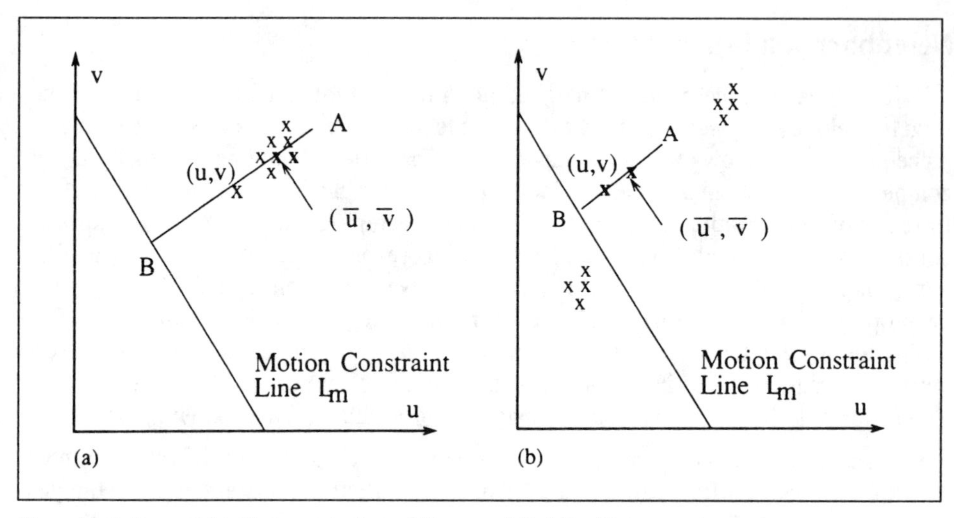

Figure 4.3. A graphical interpretation of Horn and Schunck's procedure.

This solution lends itself to the following interpretation. In steady state, the velocity (u, v) at a given pixel (see Figure 4.3a) lies on the line $\overline{AB}$ in uv space that is perpendicular to its motion constraint line L_m (given by Equation 3.4) and that passes through the point $(\bar{u}, \bar{v})$ corresponding to the average velocity in the local neighborhood. The same holds during any iteration, except that (u, v) corresponds to the current iteration and the average $(\bar{u}, \bar{v})$ is taken from the previous iteration. Thus, during any iteration, the pixel under consideration is assigned a velocity that lies somewhere on the line $\overline{AB}$, depending on the value of the smoothing parameter α. This is a reasonable procedure for the situation depicted in Figure 4.3a, where velocities of all the neighboring pixels are similar and are clustered together in uv space. This situation corresponds to a homogeneous region.

In the vicinity of a motion boundary, however, the situation is different, as depicted in Figure 4.3b. The neighborhood straddles the boundary, and velocities of various pixels in the neighborhood form two clusters in uv space. In Horn and Schunck's technique, the pixel under consideration is assigned a velocity that lies somewhere between the average velocities on the two sides of the boundary. In effect, the flow field is blurred in the vicinity of the motion boundary. To get the correct flow field, however, the pixel under consideration should be assigned a velocity that is similar to the average velocity on one side of the boundary or the other. That is, the pixel under consideration should be *binned* into one of the two locally consistent neighborhoods. Furthermore, it should be possible to do so without explicitly finding the motion boundaries in advance. In other words, the binning process should, as a side effect, produce motion boundaries.

This leads to an important observation. In smoothing-based techniques such as that of Horn and Schunck, only the mean of the neighborhood velocities $(\bar{u}, \bar{v})$ is used to represent the "opinion" of the neighborhood. Figure 4.4 shows the typical velocity

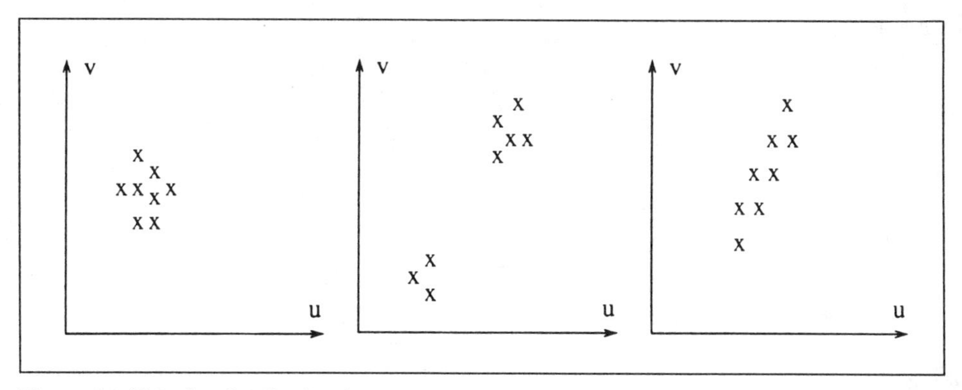

Figure 4.4. Velocity distribution for some representative neighborhoods: (a) uniform region, (b) region boundary, (c) gradual depth change.

distribution for some representative neighborhoods in *uv* space. It is apparent that these neighborhoods have the same mean. However, intuition suggests that they should have very different opinions. I suggest that the distribution of neighborhood velocities in *uv* space be used as a whole (instead of just the mean) during velocity propagation. In Chapter 5, I will show that this distribution can be interpreted to give a representation of local velocity exactly like conservation information:

- an estimate along with a covariance matrix, or
- two linear constraints, each with a confidence measure.

The second representation was used earlier by Scott [135]. In this representation, the two constraints can be thought of as the neighborhood constraints discussed in the previous chapter.

Conclusion

In this chapter, I have shown that the velocity information that can be derived from small spatiotemporal neighborhoods in time-varying imagery is inexact. Furthermore, it can be classified into two categories: conservation information and neighborhood information. In the next chapter, I show the details of a framework that computes optic flow by recovering and combining the two types of information using estimation-theoretic techniques.

Chapter 5

Estimation-Theoretic Framework: Computational Details

This chapter gives a detailed description of the computational steps that underlie the new framework outlined in Chapter 4. As described earlier, optic-flow estimation involves two functional steps: recovering conservation information and propagating it using neighborhood information. In this chapter, I discuss these two steps in view of the new framework. First, I show how to recover conservation information in the representations that were discussed, albeit cursorily, in the previous chapter. To simplify the presentation, I use the correlation-based approach. (In Chapter 8, I show that we can use any of the other two basic approaches as well to recover conservation information in the required representation, and I give details of the procedures.) Next, I discuss the procedure for recovering neighborhood information. I also show that velocity propagation can be posed as a problem of combining conservation information and neighborhood information statistically. I present an iterative solution to this problem. Then, I show two algorithms based on this framework. Finally, I give some concluding remarks and set the stage for testing the framework experimentally in Chapters 6 and 7.

Conservation information

I will use the correlation-based approach to illustrate the procedure for recovering conservation information. Unless specified otherwise, the discussion pertains to recovering this information for a given pixel in the imagery by using a small spatiotemporal neighborhood around it. After a brief qualitative description of the overall procedure, I present the quantitative details. Then, I discuss various implementation issues, such as extension of the framework to multiple images, selection of various window sizes, and normalization parameters. Finally, I discuss the relationship of this procedure with some other existing procedures.

A qualitative description. Estimating optic flow using the correlation-based approach [80] essentially involves an explicit search for the best match for a given pixel of an image in subsequent images of the sequence. Typically, such a search is conducted

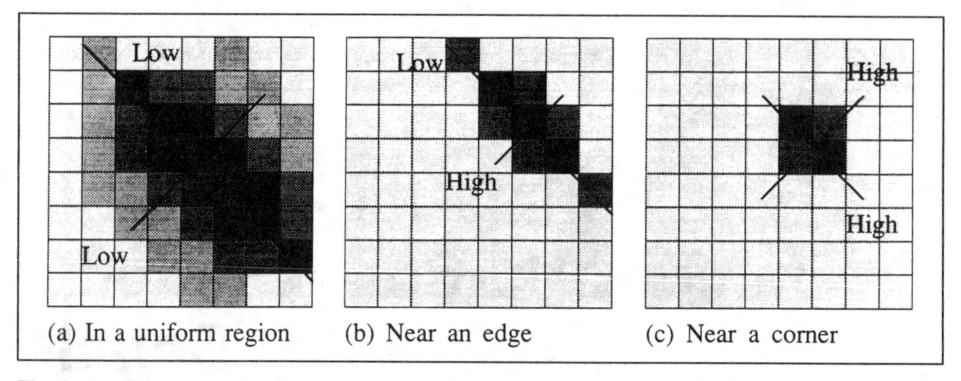

Figure 5.1. Response distribution over the search window for some representative examples: (a) in a uniform region, (b) near an edge, (c) near a corner. The darker the pixel, the higher the response. High and low refer to confidence measures associated with the principal axes.

using the procedure given below, which is based on the assumption of conservation of local spatial distribution of image intensity.

We can assume that some function of image intensity, rather than intensity itself, is conserved as time progresses. For example, Anandan [80] used conservation of the local spatial distribution of band-pass filtered image intensity. The discussion in this section is applicable, irrespective of the choice of the conserved property. (I discuss the issue of selection of conserved property later in this section.)

Here is the procedure:

1. A small window $\mathcal{W}_p$ is formed around the pixel under consideration $\mathcal{P}$ at location (x, y) in the first image. (For simplicity, I use only two images to illustrate the basic procedure and discuss its extension to multiple frames later in this section.)
2. A search area $\mathcal{W}_s$ is established a priori, around location (x, y) in the next image. The extent of this search area is determined on the basis of prior knowledge about the maximum possible image displacement.
3. The correlation between $\mathcal{W}_p$ and each corresponding window in $\mathcal{W}_s$ is then computed, thus giving a *response*—that is, a matching strength—at each pixel in the search window $\mathcal{W}_s$.

Most of the techniques suggested in the past research typically select the pixel (in $\mathcal{W}_s$) with the highest response as the *match* for the pixel $\mathcal{P}$, but such a selection is not always possible [135]. There is, however, an alternative to selecting the best match. The search area can be visualized as covered with a "response distribution," as shown in Figure 5.1. I suggest that response distribution be interpreted as follows. Each point in the search area can be a candidate for the true match. However, a point with a small response is less likely to be the true match than is a point with a high response. Without loss of generality, we can assume that the time elapsed between two successive images is unity. Thus, each point in the search area represents a point in *uv* space. With this assumption, the response distribution could be interpreted as a frequency distribution in velocity space: The response at a point depicts the frequency of occurrence, or the likelihood, of the corresponding value of velocity.

Using this interpretation of response distribution, we can compute an *estimate* of velocity using, for instance, a weighted least-squares approach, with the frequency—that is, the response—serving as the weight. Further, the true velocity is a point in *uv* space. Its displacement from any point in the search window depicts the "error" associated with the corresponding velocity. Assuming that the errors are additive, zero-mean, and independent, we can also associate a covariance matrix with the estimate. I discuss these assumptions and the computational steps in the next subsection.

Alternatively, we can characterize the information contained in the response distribution by principal-axis decomposition [83, 145]. The normalized moment of inertia about each of the two principal axes thus obtained is inversely related to the confidence in assuming that the matching pixel lies on that axis. This can be observed for some representative cases, as shown in Figure 5.1. In the regions where the intensity distribution is uniform (Figure 5.1a), there will be a high response almost everywhere in the search window $\mathcal{W}_s$. This will result in a high normalized moment of inertia about each principal axis, and thus a low confidence in assuming that the matching pixel lies along any one of the two axes.

Similarly, in the vicinity of an intensity edge, the pixels with a high response will be oriented along a line in the search window $\mathcal{W}_s$ (Figure 5.1b). This will result in a very high moment of inertia about the minor principal axis and a very low moment of inertia about the major principal axis. Consequently, there will be a low confidence in assuming that the matching pixel lies on the minor axis and a high confidence in assuming that the matching pixel lies on the major axis. This observation is consistent with the aperture problem.

Finally, with an intensity corner (Figure 5.1c), the strong matches are concentrated around a single point in the search window $\mathcal{W}_s$, giving a low moment of inertia about each principal axis. As a result, there is a high confidence in assuming that the matching pixel lies on both principal axes, that is, on their intersection.

Once again, we can assume without loss of generality that the time elapsed between two successive images is unity. Thus, the principal axes of response distribution give two orthogonal linear constraints on velocity, and the reciprocals of the normalized moments of inertia give the corresponding confidence measures. In other words, we obtain conservation information in the form of conservation constraints and their confidence measures.

An essential step in the procedure described above is computation of response distribution. Several *match measures* between two intensity patterns have been proposed in the literature. The most commonly used ones are direct correlation, mean normalized correlation, variance normalized correlation, sum of squared differences (SSD), and sum of absolute differences. Burt, Yen, and Xu [81] and Anandan [80] provide a comparative survey of the various measures. Anandan chose SSD as a measure of match in his technique for two reasons: It is always positive, and it is computationally less expensive than other measures. In accordance with Anandan's reasoning, I also have chosen SSD as the match measure. SSD essentially depicts the dissimilarity between two given patterns—the higher the SSD, the less similar the two patterns. In other words, SSD measures the error between the "observed" pattern (a subwindow of $\mathcal{W}_s$) and the "predicted" pattern $\mathcal{W}_p$, thus giving an error distribution over the search area $\mathcal{W}_s$.

I refer to $\mathcal{W}_p$ as the predicted pattern for the following reason. The assumption of conservation of local distribution of intensity implies that if the intensity distribution around a location (x, y) in an image is given by $\mathcal{W}_p$, the intensity distribution around a location $(x + \delta x, y + \delta y)$ in the subsequent image ($(\delta x, \delta y)$ being the actual displacement) must also be $\mathcal{W}_p$. That is, the assumption lets us predict that the intensity pattern around the displaced position of a pixel will be $\mathcal{W}_p$. Under ideal conditions, the error between the predicted and the observed intensity patterns will be minimum—that is, zero, at $(x + \delta x, y + \delta y)$.

The error distribution obtained from SSD must be converted to a response distribution with a maximum at the location of the best match. There are several ways to do this, such as taking the reciprocal of the error or defining a function that is exponential in the negative of the error. I will discuss a method to convert error distribution into response distribution in the next subsection.

The qualitative description given above demonstrates that there are three essential steps underlying the computation of conservation information using the correlation-based approach. They are

1. select the conserved quantity and derive it from intensity imagery,
2. compute error distribution and response distribution over the search area in the velocity space, and
3. interpret response distribution (that is, *either* compute an estimate of velocity along with a covariance matrix *or* perform principal-axis decomposition of response distribution to compute two conservation constraints and their confidences).

I will now describe the quantitative details for performing these three steps. (It will be apparent in Chapter 8 that the procedures to recover conservation information using the gradient-based approach and the spatiotemporal energy-based approach also use the same three steps.) I use a five-frame image sequence as a running example to illustrate the various steps. (Figures 5.2a and 5.2b show the first and fifth frames.) These frames were digitized from a home video. In this segment, the video camera translated upward and to the left, and the baby was (almost) stationary. Since the background is quite flat, we expect the optic-flow field to be zero in the background and nonzero on the baby (with a general direction downward and to the right). The ground-truth optic flow is not known. I will use the region around the point P in the first image to illustrate some essential computational steps.

Quantitative details. The qualitative description given in the previous subsection assumes that image intensity is conserved over time. This assumption is error prone because of noise and digitization effects. I discuss below the other possible choices for the conserved quantity. I then discuss the procedure for computing error and response distribution and for interpreting response distribution.

Selecting the conserved quantity and computing it from intensity imagery. The review given in Chapter 3 shows that various functions of intensity, when chosen as the conserved property, give more robust estimates of optic flow. Some of the possible choices are discussed below:

- Anandan [80] used the local spatial distribution of band-pass filtered image intensity in a correlation-based technique. It was shown to be better than the image intensity itself or the low-pass filtered image intensity.
- Buxton and Buxton [66] used the spatiotemporal d'Alembertian of Gaussian-smoothed intensity in a gradient-based technique. Their justification was that this quantity varies approximately linearly in the vicinity of its zero crossings. Hence, a first-order gradient-based model could be used to recover normal optic flow. However, even in regions away from zero crossings, the spatiotemporal d'Alembertian of Gaussian-smoothed intensity behaves very similar to band-pass filtered intensity, except for the effects of temporal smoothing. Hence, it can be used in a correlation-based approach as well.

For computational simplicity, I use the local spatial distribution of band-pass filtered image intensity in the scheme discussed below. I refer to the band-pass filtered image as just the "image" for brevity. I implement band-pass filtering by a difference-of-Gaussians operation using the 5×5 masks suggested by Burt [136]. (For illustration, Figure 5.5a below shows the band-pass filtered image corresponding to the image of Figure 5.2a.)

Computing error distribution and response distribution in velocity space. As discussed in the qualitative description, I use sum of squared differences (SSD) as the error function. For simplicity, I use only two images, I_1 and I_2, separated by one time unit, in the procedure discussed below. Later, I discuss extension of this framework to multiple images.

For each pixel $\mathcal{P}(x, y)$ at location (x, y) in the image I_1, we form a window $\mathcal{W}_p$ of size $(2n + 1) \times (2n + 1)$ around the pixel. We establish the search window $\mathcal{W}_s$ of size $(2N + 1) \times (2N + 1)$ around the pixel at location (x, y) in the image I_2. The $(2N + 1) \times (2N + 1)$ sample of error distribution, whose elements represent the dissimilarity between $\mathcal{W}_p$ and a $(2n + 1) \times (2n + 1)$ window around *each* pixel in $\mathcal{W}_s$, is computed as

$$\mathcal{E}_c(u, v) = \sum_{i=-n}^{n} \sum_{j=-n}^{n} (I_1(x+i, y+j) - I_2(x+u+i, y+v+j))^2$$
$$-N \le u, v \le +N \qquad \textbf{(5.1)}$$

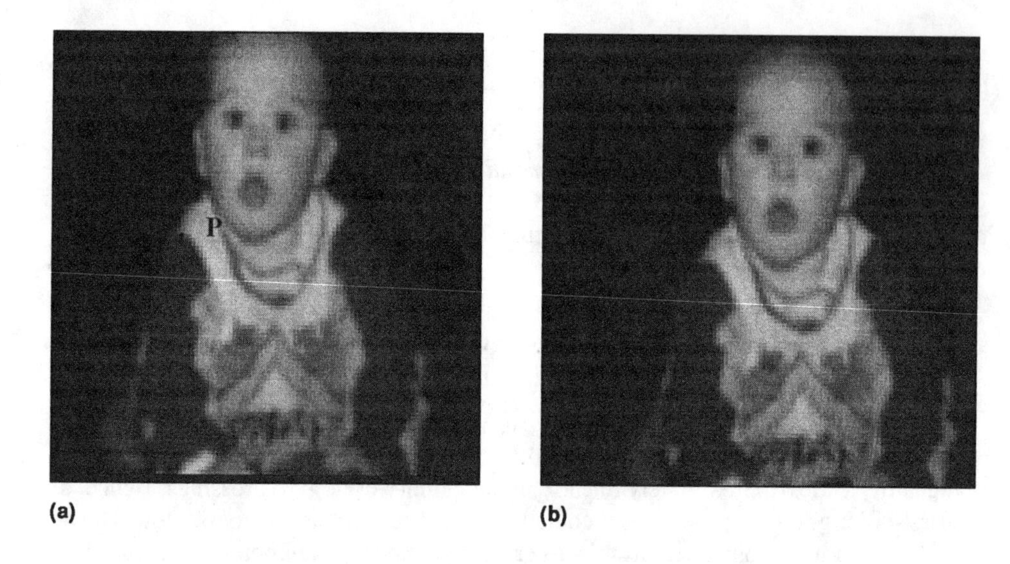

(a) (b)

Figure 5.2. The imagery used in the running example: (a) first frame, (b) fifth frame.

In this expression, $I_1(x, y)$ and $I_2(x, y)$ refer to the intensities at location (x, y) in images I_1 and I_2, respectively.

The $(2N + 1) \times (2N + 1)$ sample of response distribution, whose elements represent the similarity between $\mathcal{W}_p$ and a $(2n + 1) \times (2n + 1)$ window around each pixel in $\mathcal{W}_s$, is computed as the exponential of the negated error. That is,

$$\mathcal{R}_c(u, v) = e^{-k\left[\sum_{i=-n}^{n}\sum_{j=-n}^{n}(I_1(x+i, y+j) - I_2(x+u+i, y+v+j))^2\right]} \tag{5.2}$$

where both u and v vary between $-N$ and $+N$.

The choice of an exponential function for converting error distribution into response distribution is based primarily on computational considerations. First, it is well-behaved when error approaches zero. A function that uses, for instance, the reciprocal of error, tends to infinity as error approaches zero. Therefore, it is computationally harder to manipulate. Secondly, the response obtained with an exponential function varies continuously between zero and unity over the entire range of error.

We have to establish the parameters n, N, and k to compute response distribution. The criteria for selecting the values of these parameters are discussed later in this section. For illustration, Figure 5.3a shows the intensity pattern in a 3×3 neighborhood $\mathcal{W}_p$ of a point P in the band-pass filtered version of Figure 5.2a. Figure 3b shows a 7×7 neighborhood around the corresponding point in the consecutive image. Using $n = 1$, $N = 2$, and $k = 10^{-4}$, the 5×5 response distribution can be computed for the point P, as shown in Figure 5.3c.

20	30	121
16	27	119
81	79	111

(a)

4	6	8	9	12	19	22
5	6	8	10	12	20	22
11	11	10	23	32	124	126
14	15	16	18	26	121	121
58	60	65	82	83	116	117
98	100	105	110	112	123	124
100	104	110	110	112	124	124

(b)

9	12	18	63	82
11	17	26	154	78
67	107	162	987	143
171	194	180	398	96
157	153	106	87	39

(c)

Figure 5.3. An illustration of response-distribution computation: (a) the 3×3 window $\mathcal{W}_p$, (b) the 5×5 window $\mathcal{W}_s$ (a one-pixel-wide annulus around this window is shown because it is required to compute the correlation at boundary pixels), (c) the 5×5 response distribution ($\times 10^3$).

Interpreting response distribution. A very important step in the current framework is interpreting the response distribution. At this step the framework incorporates estimation-theoretic techniques. Here, the response distribution obtained in the previous step is interpreted to get an initial estimate of velocity of the pixel under consideration.

Each point (u, v) in the search window (in uv space) is a candidate for the true velocity. However, a point with a small response is less likely to represent the true velocity than is a point with a high response. Thus, response distribution could be interpreted as a frequency distribution in velocity space—the response at a point depicting the frequency of occurrence, or the likelihood, of the corresponding value of velocity. This interpretation lets us use a variety of estimation-theoretic techniques to compute velocity and associate a notion of confidence with it.

Specifically, the quantity we are trying to compute is the true velocity (u_t, v_t). With the interpretation given above, we know the frequency of occurrence $\mathcal{R}_c(u, v)$ of various values of velocity $(u, v) = (u_t, v_t) + (e_u, e_v)$ over the search area. The quantity (e_u, e_v) is the error associated with the point (u, v), that is, its deviation from the true velocity. We can obtain an *estimate* of the true velocity using a weighted least-squares approach [144]. This estimate, denoted by $U_{cc} = (u_{cc}, v_{cc})$, is given by

$$u_{cc} = \frac{\sum_u \sum_v \mathcal{R}_c(u, v)u}{\sum_u \sum_v \mathcal{R}_c(u, v)}$$

$$v_{cc} = \frac{\sum_u \sum_v \mathcal{R}_c(u, v)v}{\sum_u \sum_v \mathcal{R}_c(u, v)} \tag{5.3}$$

where the summation is carried out over $-N \leq u, v \leq +N$. For example, the estimate corresponding to the response distribution of Figure 2.3c is (0.35, –0.42). We can also think of this estimate as the initial estimate that would serve as input to the velocity-propagation procedure, making use of neighborhood information.

So far, nothing has been said regarding the errors (e_u, e_v). By interpreting (e_u, e_v) as the deviation of the point (u, v) from the true velocity, as discussed above, we have imposed an *additive error model* [144]. Further, if we make the following assumptions, we can associate a notion of a covariance matrix with the estimate.

- *The errors have a zero mean.* A sufficient (but not necessary) condition for the errors to have a zero mean is that the response distribution is symmetrical around the true velocity. We can satisfy this condition in regions of smoothly varying intensity if we use a hierarchical implementation [80]. In this implementation, an approximate location of the point in uv space that corresponds to the true velocity is known from lower levels of resolution. A symmetrical search window is established around this point to refine the velocity estimate. With this construction, the search window is approximately symmetrical about the true velocity. Also, since the region is of smoothly varying intensity, the response distribution is unimodal, with its mode approximately at the true velocity. This results in an approximately zero-mean error. In a single-resolution implementation, the search window is symmetrical about the origin of uv space, not about the true velocity. This does cause the mean error to deviate from zero. Similarly, if the region under consideration is textured, it may lead to multiple modes within the search window, only one of which is at the true velocity. This also causes the mean error to deviate from zero.

- *The errors are independent.* This assumption is almost always violated because the response at various pixels in the search window is obtained using overlapping correlation windows. Also, in a hierarchical implementation, an error incurred at a pixel at a coarse resolution is transmitted to several neighboring pixels in finer levels. This causes the errors to be dependent. However, as the experiments in the next chapter suggest, the violation of this assumption does not seem to have a major effect on optic-flow estimates.

With these assumptions, we can associate the following covariance matrix with the estimate given above:

$$S_{cc} = \begin{pmatrix} \dfrac{\sum_u \sum_v \mathcal{R}_c(u,v)(u-u_{cc})^2}{\sum_u \sum_v \mathcal{R}_c(u,v)} & \dfrac{\sum_u \sum_v \mathcal{R}_c(u,v)(u-u_{cc})(v-v_{cc})}{\sum_u \sum_v \mathcal{R}_c(u,v)} \\ \dfrac{\sum_u \sum_v \mathcal{R}_c(u,v)(u-u_{cc})(v-v_{cc})}{\sum_u \sum_v \mathcal{R}_c(u,v)} & \dfrac{\sum_u \sum_v \mathcal{R}_c(u,v)(v-v_{cc})^2}{\sum_u \sum_v \mathcal{R}_c(u,v)} \end{pmatrix} \quad \mathbf{(5.4)}$$

Again, the summation is carried out over $-N \le u, v \le +N$.

Alternatively, we could quantify the response distribution using a pair of mutually orthogonal conservation constraints, each with a confidence measure associated with it. These constraints, being the principal axes of response distribution, can be computed simply as the eigenvectors of the covariance matrix defined above, and we can write them in the following form [105, 135]:

$$\begin{aligned} \mathcal{L}_{1c} &: a_{1c}u + b_{1c}v + c_{1c} = 0 \\ \mathcal{L}_{2c} &: a_{2c}u + b_{2c}v + c_{2c} = 0 \end{aligned} \quad \mathbf{(5.5)}$$

The two constraints intersect at $U_{cc} = (u_{cc}, v_{cc})$. The confidence measures C_{1c} and C_{2c} associated with these constraints can easily be computed as reciprocals of the normalized moments of inertia of the response distribution about these two lines. The normalized moments of inertia are equal to the squares of the eigenvalues of the covariance matrix. For example, conservation constraints obtained from the response distribution of Figure 5.3c are $-0.55u + 0.84v + 0.54 = 0$ and $-0.84u - 0.55v + 0.07 = 0$. The confidence measures of these constraints are 1.42 and 0.56, respectively.

An assumption implicit in the preceding discussion needs explanation. In interpreting the response distribution, I have assumed that it is unimodal. That is, it does not have multiple peaks within the search window. This assumption is violated if the search window contains repeated intensity patterns. Such a situation can arise if the size of the search window is larger than the scale of intensity variations. In this case, the correct velocity corresponds to one of the peaks. On the other hand, the weighted least-squares approach used above averages out the various peaks. Therefore, the resulting estimate is incorrect. However, the spread of the distribution is large in this case (as compared with the situation where the response distribution has a single well-defined peak). Therefore, the confidence associated with the estimate will be low. Summarizing, although the procedure for interpreting the response distribution gives an incorrect estimate if the distribution is not unimodal, it does associate a low confidence with the (incorrect) estimate.

Some implementation issues. Three important issues are related to the implementation of the method described above: extension of the computational framework to work with more than two images, selection of various window sizes, and selection of the normalization parameter k used in computing the response distribution from the error distribution.

Extension to multiple images. The procedure to compute conservation information, as described in the previous subsection, uses only two images. The issue of extending this procedure to work with more than two images can be viewed from two aspects.

First, imagine a scenario where a camera views a dynamic scene and acquires the images at a fast rate. Obviously, optic flow is time-dependent and must be updated for every image. Can the basic measurement procedure that uses only two images be incorporated into a framework that updates the flow field for each new image acquired? This issue has to do with proper housekeeping during computation. It has been addressed by several previous works, such as that of Horn and Schunck [34].

Second, are there any inherent drawbacks in using just two images in the basic computational step, that is, deriving conservation information? If yes, what is the remedy? This aspect has been addressed in earlier research using the gradient-based approach, particularly the work of Buxton and Buxton [66] and Waxman [78]. In the following discussion, I address it for the case of the correlation-based approach. I show that using just two images for computation of conservation information has a limitation. It can lead to incorrect estimates of velocity in textured image regions if the scale of texture is smaller than that of various window sizes. I also show that this problem can be alleviated, under certain specific operating conditions, by using three images. The experiments described in this and subsequent chapters use this extended scheme based on three images as their basic computational step.

Imagine the one-dimensional scenario shown in Figure 5.4. Three "images" I_{-1}, I_0, and I_{+1} taken at three successive closely spaced time instants are shown in Figure 5.4a. Image I_0 has two distinctive features, one located at $x = 0$ and the other at $x = x_1$ along the space axis. Assuming that the pattern on the image is translating at some constant speed (this is a reasonable assumption over small spatiotemporal neighborhoods), these features will be located in image I_{+1} at $x = \delta x$ and $x = x_1 + \delta x$, respectively, and in image I_{-1} at $x = -\delta x$ and $x = x_1 - \delta x$, respectively. Figure 5.4a also shows the relative sizes of $\mathcal{W}_p$ and $\mathcal{W}_s$.

If correlation is performed between I_0 and I_{+1}, two peaks in the response distribution $\mathcal{R}_c^{+1}$ are obtained at $x = \delta x$ and $x = x_1 + \delta x$, respectively. The mean of $\mathcal{R}_c^{+1}$ lies at $x_1/2 + \delta x$, which is an incorrect estimate of displacement for the pixel located at $x = 0$. However, if correlation is performed between I_0 and I_{-1}, two peaks in the response distribution $\mathcal{R}_c^{-1}$ are obtained at $x = -\delta x$ and $x = x_1 - \delta x$, respectively. This is depicted in Figure 5.4b.

Suppose we "rotate" $\mathcal{R}_c^{-1}$ about $x = 0$ and add it to $\mathcal{R}_c^{+1}$. There will be three peaks in the resultant response distribution $\mathcal{R}_c$—at $\delta x - x_1$, δx, and $x_1 + \delta x$, respectively. This is shown in Figure 5.4c. The mean of this distribution is at δx, which is the correct estimate of displacement for the pixel located at $x = 0$. This scheme is truly applicable only if the sizes of the correlation window and the search window are such that the three peaks in the resultant response distribution have the spatial arrangement shown in Figure 5.4c. However, it has empirically proved to be a useful heuristic and is used in all the experiments reported here.

Selection of window sizes. The size of $\mathcal{W}_p$ is determined on the basis of how many neighbors should contribute their opinion in an estimation of the velocity of the point under consideration. Too small a neighborhood leads to noisy estimates. Too large a neighborhood tends to smooth out the estimates. Empirically, a 3×3 window appears appropriate (see Chapter 6 for experiments). The size of $\mathcal{W}_s$ can be selected in two ways. If an a priori estimate of the maximum possible displacement of a pixel is available, say N, we can choose the size of $\mathcal{W}_s$ to be $(2N + 1) \times (2N + 1)$, thus ensuring that the matching pixel lies within the search window. Most of the experiments shown in this and the subsequent chapters assume that a point cannot be displaced by more than two pixels. Hence, a 5×5 search window is appropriate.

Alternatively, a coarse-to-fine search strategy could be used [80] to decide the extent of the search window at any level of resolution. This strategy is particularly useful when the displacements to be estimated are large. I describe some experiments in the next chapter where the displacements over two successive frames range from one pixel to eight pixels. I also show how a hierarchical version of the current framework can be used to recover the flow field. This version of the framework uses a coarse-to-fine strategy to establish the extent of the search window.

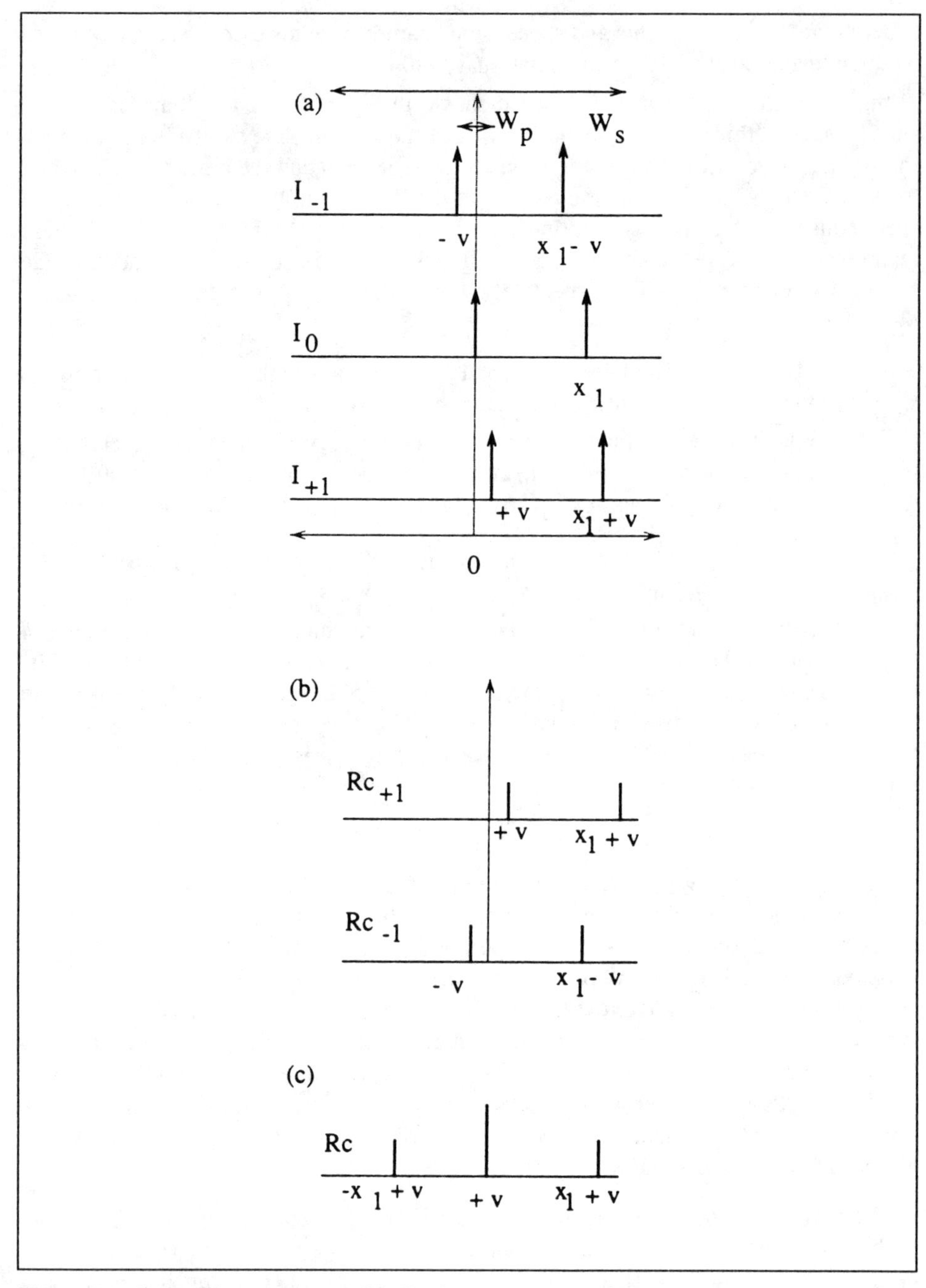

Figure 5.4. Computing response distribution using three images: (a) three consecutive images with a pair of translating features, (b) response distributions $\mathcal{R}_c^{+1}$ and $\mathcal{R}_c^{-1}$, (c) the resultant response distribution $\mathcal{R}_c$ obtained by rotating $\mathcal{R}_c^{-1}$ about 0 and adding it to $\mathcal{R}_c^{+1}$. The mean of $\mathcal{R}_c$ is at δx.

Selection of the normalization parameter. Finally, some criterion has to be established for selecting the value of the parameter k used in converting error distribution into response distribution. Empirically, I have found that as long as the response distribution is unimodal within the search window and the correct velocity does not lie close to the boundary of the search window, its (weighted least-squares) estimate is quite insensitive to the choice of k. However, if the response is too small everywhere in the search window, the divisor in the elements of the covariance matrix in Equation 5.4 tends to be very small. This leads to computational difficulties (divide-by-zero errors). To avoid this behavior, I have chosen a value of k such that the maximum response is a number close to unity.

For all the experiments in this book, I have chosen this number to be 0.95. This procedure not only raises the maximum response in the search window, but also distorts the "shape" of the response distribution. Further research needs to be done to establish the overall effect of this distortion on the final estimate of velocity and the associated covariance matrix. This procedure amounts to computing a different value of k for each pixel in the image. Further, to use three images to compute response distribution, I use the smaller of the two k's, obtained for $\mathcal{R}_c^{-1}$ and $\mathcal{R}_c^{+1}$, respectively. Over the various image sequences used in the experiments in this and the subsequent chapters, k ranges between 10^{-4} and 10^{-1}.

Figure 5.5 illustrates the estimates obtained for the running example. Three successive 128×128 images of the image sequence are used. The sizes of $\mathcal{W}_p$ and $\mathcal{W}_s$ are 3×3 and 5×5, respectively. Figure 5.5a shows the band-pass filtered version of one of the images. Figures 5.5b and 5.5c show images depicting the confidences C_{1c} and C_{2c} at each point. The velocity estimates (u_{cc}, v_{cc}) over the image are shown in Figure 5.5d. This example only illustrates the principle and does not make any claims about performance. I present a detailed experimental evaluation in Chapter 6.

Relationship with major current approaches. My approach is inspired primarily by the work of Anandan [80] and Scott [83]. There are, however, some salient differences between my approach and their approaches. The following discussion clarifies the differences.

Anandan uses SSD itself as a measure of (mis)match. He assumes that there is a well-defined minimum in the SSD surface and uses the distance of the minimum from the pixel under consideration as the estimate of displacement. He then computes the principal *curvatures* of the SSD surface at this minimum and uses them as confidence measures.

This procedure is hard to justify. If there are more than one minimum in the search window, Anandan's procedure will pick up one of them as the best match. Then, it might assign a high confidence to it, if the principal curvatures at this minimum are large. Using normalized moments of inertia, however, gives an estimate that averages out the two peaks in the response distribution and assigns a lower confidence to the estimate. This

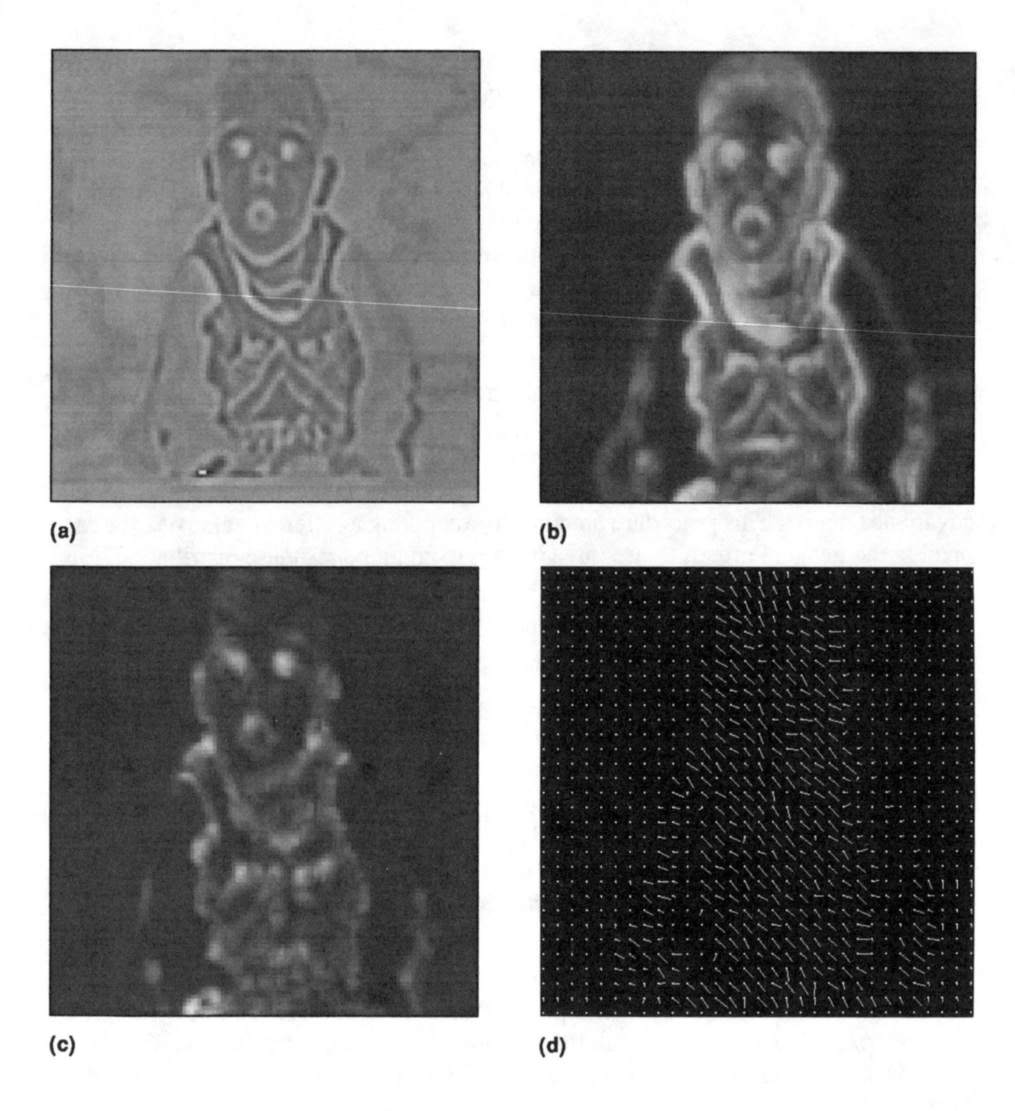

Figure 5.5. An illustration of correlation-based velocity estimates and confidence measures for the running example: (a) band-pass filtered version of one of the images, (b) confidence measure $\mathcal{C}_{1c}$, (c) confidence measure $\mathcal{C}_{2c}$, (d) estimates of velocity (u_{cc}, v_{cc}).

is appropriate because of the high uncertainty associated with local measurement of motion in this example.

Anandan does use a coarse-to-fine search strategy that can disambiguate between two possible matches. In that case, it may be appropriate to assign a high confidence to the (correct) estimate. However, the coarse-to-fine search strategy does not always guarantee correct disambiguation.

The procedure to derive velocity estimates and confidence measures in my framework is more similar to Scott's than to Anandan's. However, it differs from Scott's procedure in the following aspects:

- *Approach used for obtaining conservation information.* Scott uses the correlation-based approach to recover conservation information. I show (in Chapter 8) that this information can be derived using the gradient-based and spatiotemporal energy-based approaches also. The information obtained is in the same form—an estimate accompanied by a covariance matrix—irrespective of the approach used. I show (in Chapter 9) that conservation information can also be obtained by integrating the three basic approaches.

- *Representation of velocity information.* Whereas Scott uses the principal-axis-based representation for velocity, I use the estimate-covariance-based representation. Even though the two representations are conceptually identical, the estimate-covariance representation is computationally easy to use. Also, it allows the problem of velocity computation to be viewed as a parameter-estimation problem that can be solved using standard estimation-theoretic techniques. This will be demonstrated in the next section. Further, it allows unification and integration of various existing approaches for optic-flow computation. I discuss the issues of unification and integration in detail in Chapters 8 and 9.

- *Selection of the conserved property.* Scott uses the local spatial distribution of image intensity as the conserved property. As discussed in Chapter 3, this assumption is error prone. On the other hand, flow estimates based on conservation of the local spatial distribution of band-pass filtered intensity are more robust. For this reason, I use band-pass filtering as a preprocessing stage to the computation of the response distribution.

- *Computation of the response distribution.* Scott converts error distribution into response distribution by using a rational function given by

$$\frac{k_1}{k_2 + k_3 E_c}$$

  On the other hand, I use an exponential which is better behaved in terms of its range and its characteristics when error approaches zero. This offers a considerable computational advantage.

- *Range of velocities.* Scott uses the time-varying imagery at its original spatial resolution. This limits to the size of the search window the range of velocities that can be estimated. I show in a later section that a multiresolution strategy can be embedded in the framework discussed above. This greatly extends the range of velocities that can be estimated.

Step 2: Neighborhood information

The objective of the second step in optic-flow recovery is to propagate velocity by using neighborhood information. In this section, I discuss how to obtain neighborhood information in the form discussed in the previous chapter and how to use it with conservation information to recover optic flow.

A qualitative description. Assume for a moment that the velocity of each pixel in a small neighborhood around the pixel under consideration is known. We could plot these velocities as points in *uv* space giving a neighborhood velocity distribution. Some typical distributions were shown in Figure 4.4, and are reproduced in Figure 5.6. What can we say about the velocity of the central pixel (which is unknown)? Barring the case where the central pixel lies in the vicinity of a motion boundary, it is reasonable to assume that it is "similar" to velocities of the neighboring pixels. (In the next subsection, I will discuss the performance of the method presented here for the pixels lying in the vicinity of a motion boundary.)

In statistical terms, the velocity of each point in the neighborhood can be thought of as a measurement of the velocity of the central pixel. It is reasonable to assume that all of these measurements are not equally reliable—they must be weighted differently if used to compute an estimate of velocity of the central pixel. I weight the velocities of various pixels in the neighborhood according to their distance from the central pixel—the larger the distance, the smaller the weight. Specifically, I use a Gaussian mask.

This construction makes neighborhood information available in the form of a set of weighted measurements of the true velocity. We can use this information to compute a weighted least-squares estimate of velocity. Furthermore, with the assumptions of additive, zero-mean, and independent errors in the measurements, we can associate a covariance matrix with this estimate. The estimate and the covariance matrix thus obtained serve as the "opinion" of the neighborhood regarding the velocity of the central pixel (as opposed to those obtained from conservation information that reflect the central pixel's own opinion). Alternatively, as shown in Figure 5.6, we can perform a principal-axis decomposition of neighborhood velocity distribution to compute two mutually orthogonal linear constraints and the confidence measures associated with them. In view of the terminology proposed in Chapter 3, these constraints are the neighborhood constraints.

Now we have two estimates of velocity, U_{cc} and $\overline{U}$. These come from conservation and neighborhood information, respectively, and each has a covariance matrix. We can compute an estimate of velocity that takes both conservation information and neighborhood information into account. Since this estimate is a point in *uv* space, its distance from $\overline{U}$, weighted appropriately by the corresponding covariance matrix, represents the error in satisfying neighborhood information. I refer to this error as *neighborhood error*. Similarly, the distance of this point from U_{cc}, weighted appropriately, represents the error in satisfying conservation information. I refer to this error as *conservation error*. Computing the velocity estimate, therefore, amounts to finding a point in *uv* space that

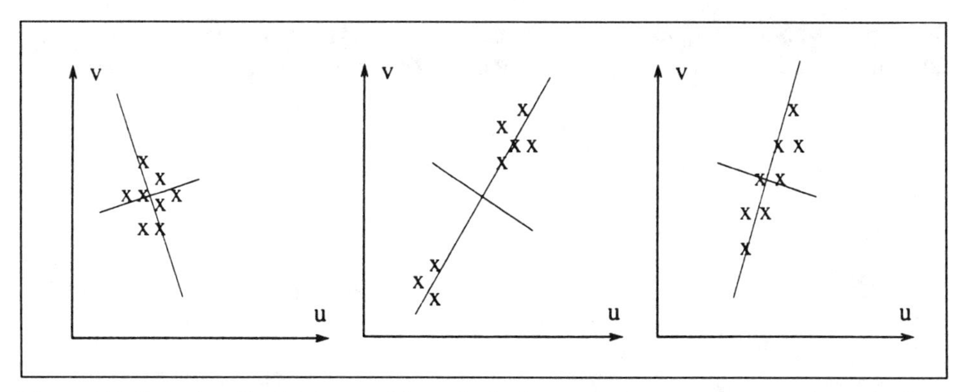

Figure 5.6. Velocity distribution and its principal axes for some representative neighborhoods: (a) uniform region, (b) region boundary, (c) gradual depth change.

minimizes the sum of neighborhood error and conservation error. Because we lack any additional knowledge, we weight the two errors equally. In the next subsection, I pose the problem of estimating velocity in this fashion in statistical terms and show an iterative solution to the problem.

I have assumed all along in this qualitative description that neighborhood velocities are correctly known in advance from an independent source. This is not true in reality, so the solution is iterative. I explain the details of the procedure in the next subsection.

Quantitative details. First, I show how to interpret neighborhood information in statistical terms to compute an estimate of velocity of the central pixel (in the opinion of the neighborhood) along with a covariance matrix. Second, I discuss how to formulate conservation error and neighborhood error in statistical terms. Then, I use these formulations to pose the problem of optic-flow recovery as that of minimizing the mean value of overall error and give an iterative solution to this problem. I give a geometric interpretation of the iterative solution and show that it amounts to performing a statistical combination of conservation information and neighborhood information at each iteration. Finally, I use the geometric interpretation to show that the iterative update scheme does not blur the flow field at certain types of motion boundaries.

Interpreting neighborhood information. Once again, assume that the velocities of all the pixels in a small window (of size $(2w + 1) \times (2w + 1)$) around the pixel under consideration are available from an independent source. We can treat them as measurements of the velocity of the central pixel. Each velocity is mapped onto a point in *uv* space and is assigned a weight that depends on the distance of the corresponding pixel from the center of the window. (In a later section, I discuss the actual weight assignment.) Let the $(2w + 1)^2$ points thus obtained in *uv* space be denoted by (u_i, v_i), where $1 \leq i \leq (2w + 1)^2$. Let the weight assigned to the point (u_i, v_i) be given by $\mathcal{R}_n(u_i, v_i)$. Assuming that there are no outliers, these measurements can be used to compute an estimate based on weighted least squares, $\overline{U} = (\overline{u}, \overline{v})$, of the velocity of the central pixel as follows:

$$\overline{u} = \frac{\sum_u \sum_v \mathcal{R}_n(u, v)u}{\sum_u \sum_v \mathcal{R}_n(u, v)}$$
$$\overline{v} = \frac{\sum_u \sum_v \mathcal{R}_n(u, v)v}{\sum_u \sum_v \mathcal{R}_n(u, v)} \tag{5.6}$$

It is apparent that the velocity measurements (u_i, v_i) are erroneous. That is, they are not all exactly equal to the true velocity of the central pixel. Under the assumptions of additive, zero-mean, and independent errors, we can associate the following covariance matrix with the estimate [144]:

$$\mathcal{S}_n = \begin{pmatrix} \frac{\sum_i \mathcal{R}_n(u_i, v_i)(u_i - \overline{u})^2}{\sum_i \mathcal{R}_n(u_i, v_i)} & \frac{\sum_i \mathcal{R}_n(u_i, v_i)(u_i - \overline{u})(v_i - \overline{v})}{\sum_i \mathcal{R}_n(u_i, v_i)} \\ \frac{\sum_i \mathcal{R}_n(u_i, v_i)(u_i - \overline{u})(v_i - \overline{v})}{\sum_i \mathcal{R}_n(u_i, v_i)} & \frac{\sum_i \mathcal{R}_n(u_i, v_i)(v_i - \overline{v})^2}{\sum_i \mathcal{R}_n(u_i, v_i)} \end{pmatrix} \tag{5.7}$$

where the summation is carried out over $1 \leq i \leq (2w + 1)^2$.

The assumption of independence is much harder to justify for neighborhood information than for conservation information. As I discuss later, the neighborhood information is, in fact, derived from conservation information. If conservation information is consistently erroneous in a given neighborhood, the neighborhood errors will be strongly correlated. Similarly, the assumption of a zero mean is clearly violated in the vicinity of a motion boundary. Further research needs to be done to ascertain the validity of these assumptions and to quantify the effects of violating them on the final estimate of velocity.

An alternative approach is to represent neighborhood information with a pair of mutually orthogonal linear constraints, referred to as neighborhood constraints, each with a confidence measure. These constraints are the principal axes of the weighted neighborhood velocity distribution. The principal axes are simply the eigenvectors of the covariance matrix $\mathcal{S}_n$ defined above and can be written in the following form [105, 135]:

$$\mathcal{L}_{1n} : a_{1n}u + b_{1n}v + c_{1n} = 0$$
$$\mathcal{L}_{2n} : a_{2n}u + b_{2n}v + c_{2n} = 0 \qquad (5.8)$$

The two constraints intersect at $\overline{U} = (\overline{u}, \overline{v})$. The confidence measures C_{1n} and C_{2n} associated with them can easily be computed as reciprocals of the normalized moments of inertia of response distribution about these two lines. The normalized moments of inertia are equal to the squares of the eigenvalues of the covariance matrix.

Neighborhood error and conservation error. I will use the estimate-covariance representation to write the expressions for neighborhood error and conservation error. I use U, a 2×1 vector, to denote the correct velocity of the pixel under consideration. Neighborhood error, as defined in the earlier qualitative discussion, is a quadratic form commonly used in estimation theory [144] and is given by

$$(U - \overline{U})^T S_n^{-1} (U - \overline{U}) \qquad (5.9)$$

where $\overline{U}$ is the 2×1 vector and S_n is the 2×2 matrix defined in Equations 5.6 and 5.7, respectively. The superscript T denotes matrix transposition and the superscript -1 denotes matrix inversion. Similarly, conservation error has the following quadratic form:

$$(U - U_{cc})^T S_{cc}^{-1} (U - U_{cc}) \qquad (5.10)$$

where U_{cc} is the 2×1 vector and S_{cc} is the 2×2 matrix, as defined in Equations 5.3 and 5.4, respectively.

The minimization problem. The sum of conservation error and neighborhood error represents the squared error in the velocity estimate U. We can obtain an estimate of velocity that, under the assumptions of additive, zero-mean, and independent errors, minimizes the mean error over the visual field. That is

$$\iint [(U - \overline{U})^T S_n^{-1} (U - \overline{U}) + (U - U_{cc})^T S_{cc}^{-1} (U - U_{cc})] \, dxdy = \text{Minimum} \qquad (5.11)$$

A calculus of variations [146] can be used to derive the estimate. Let an operator ∇_U be defined as follows:

$$\nabla_U = \begin{pmatrix} \frac{\partial}{\partial u} \\ \frac{\partial}{\partial v} \end{pmatrix} \tag{5.12}$$

The condition for minimum mean-squared error can be written as

$$\nabla_U \left[\iint [(U - U_{cc})^T S_{cc}^{-1} (U - U_{cc}) + (U - \overline{U})^T S_n^{-1} (U - \overline{U})] \, dxdy \right] = 0 \tag{5.13}$$

which gives [144]

$$S_{cc}^{-1} (U - U_{cc}) + S_n^{-1} (U - \overline{U}) = 0 \tag{5.14}$$

In this equation, U_{cc} and S_{cc} are derived directly from the underlying intensity pattern in the image. Therefore, they are known (and fixed) for each pixel. On the other hand, $\overline{U}$ and S_n are derived from the assumption that the velocity of each pixel in the neighborhood is known in advance from an independent source (see the earlier qualitative description). In practice, this assumption is invalid. Hence, $\overline{U}$ and S_n are unknown and the velocity U cannot be derived directly from Equation 5.14. However, Equation 5.14 is available at all the pixels in any given neighborhood in the image. Further, $\overline{U}$ is a linear function of the velocities in the neighborhood, and S_n is a quadratic function of the velocities in the neighborhood. Therefore, computing U amounts to solving a system of coupled nonlinear equations.

This is a computationally difficult problem. However, if the conditions listed below are satisfied, we essentially have a system of coupled linear equations that can be solved by an iterative technique such as the Gauss-Siedel relaxation algorithm [147]. The iterative solution can be written as [147]

$$U^{k+1} = \left[S_{cc}^{-1} + S_n^{-1} \right]^{-1} \left[S_{cc}^{-1} U_{cc} + S_n^{-1} \overline{U}^k \right]$$
$$U^0 = U_{cc} \tag{5.15}$$

The iterative solution of Equation 5.15 cannot be used directly for optic-flow estimation because it breaks down if one of the matrices is singular (or almost singular) and cannot be inverted reliably. I discuss a practical way to implement the iterative procedure in a later section on implementation issues.

For Equation 5.14 to represent a system of coupled *linear* equations, S_n^{-1} must be a constant and must be known in advance. Such is not the case here. In the current implementation, I obtain S_n from the neighborhood velocity distribution corresponding to the previous iteration. However, I have found empirically that S_n^{-1} (that is, any of its elements) does not change by more than about 15 percent from the beginning to the end of the iterative procedure. This holds true particularly for the pixels that do not lie on a motion boundary.

For the iterative procedure to converge irrespective of the value of initial estimate U^0, both S_{cc}^{-1} and S_n^{-1} must be positive definite. This criterion is generally satisfied in real imagery except in pathological cases such as absolutely flat regions [148].

The choice of U_{cc} as the starting velocity for the iterative procedure is justified because it denotes the estimate that can be derived from conservation information alone. This ties well with the two-step approach to optic-flow recovery: The output of the first step, U_{cc}, serves as an input to the second step. I show below that each iteration of the iterative procedure amounts to performing an MSE-optimal combination of conservation information with neighborhood information derived from the previous iteration. (MSE stands for mean-squared error.) This can be used to give a geometric interpretation of the iterative procedure and to explain the performance of this framework at motion boundaries.

Geometric interpretation. To give a geometric interpretation of the iterative procedure, I will use the *information fusion theorem* that will be proved in Chapter 9 in a different context.

Information fusion theorem. There are n sensors, each giving a measurement X_i of an unknown quantity X with an additive, zero-mean, and independent error V_i, where X_i, X, and V_i are $p \times 1$ vectors. Each error V_i is characterized by its $p \times p$ covariance matrix S_i. The optimal estimate $\hat{X}$ for the unknown quantity X and its associated covariance matrix $\hat{S}$ are given as follows. The optimality is meant to imply that the error in the fused estimate is unbiased and is minimal in a mean-squared sense. In other words, the fused estimate is MSE-optimal:

$$\hat{X} = [S_1^{-1} + S_2^{-1} + \ldots\ldots + S_n^{-1}]^{-1} [S_1^{-1}X_1 + S_2^{-1}X_2 + \ldots\ldots + S_n^{-1}X_n]$$
$$\hat{S} = [S_1^{-1} + S_2^{-1} + \ldots\ldots + S_n^{-1}]^{-1} \qquad \textbf{(5.16)}$$

Equation 5.15 is clearly identical in its form to Equation 5.16 for $n = 2$. Thus, Equation 5.15 can be interpreted as follows. At each iteration, conservation information and neighborhood information act as two sensors giving U_{cc} and $\overline{U}^k$ as their respective measurements of the velocity U, with error characteristics given by S_{cc} and S_n, respectively. The updated velocity U^{k+1} is the MSE-optimal combination of the two measurements, as suggested by the theorem. The covariance associated with the combined estimate U^{k+1} is given by $[S_{cc}^{-1} + S_n^{-1}]^{-1}$. Given that both sensors measure the same mean and the principal axes of the two ellipses are aligned, the resultant covariance is smaller

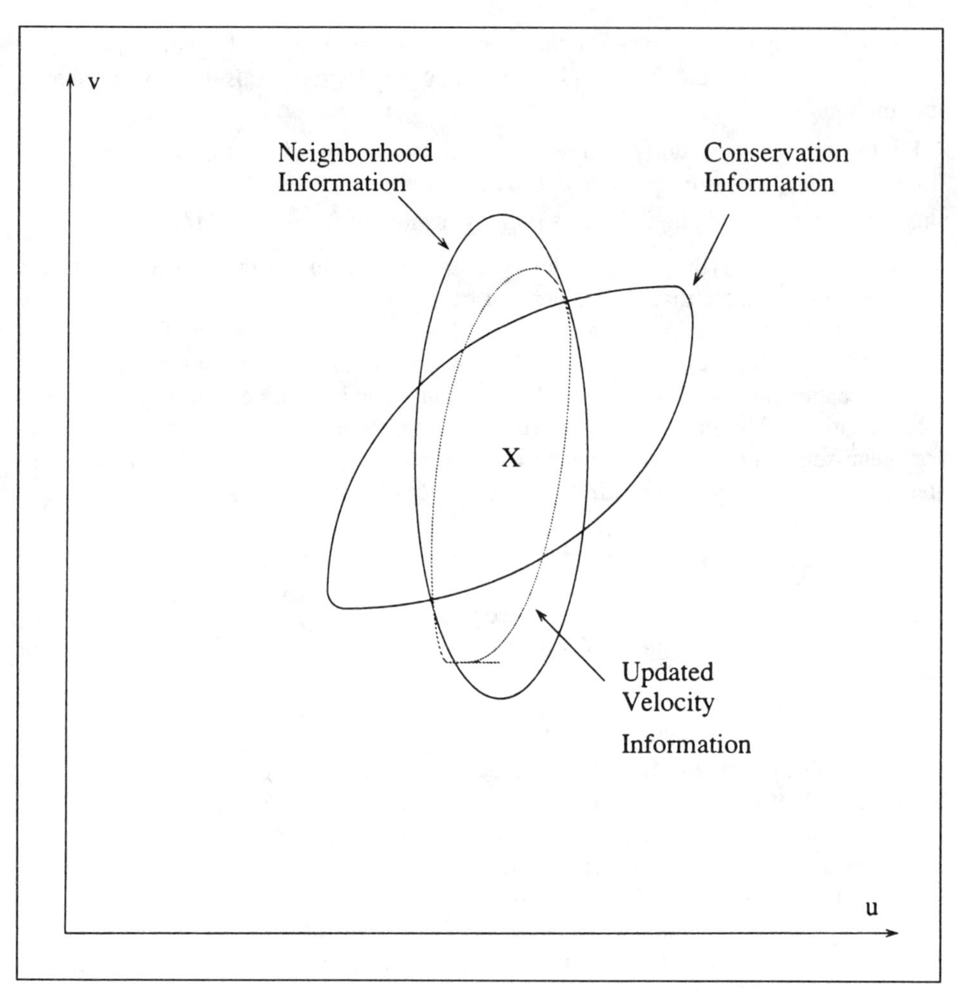

Figure 5.7. A geometric interpretation of the iterative procedure.

than the individual covariances. This can be geometrically interpreted, as shown in Figure 5.7. Conservation information and neighborhood information at any iteration are represented by estimates accompanied by regions of uncertainty. At each iteration, the updated velocity is represented by an estimate that lies somewhere between the conservation and neighborhood estimates. Furthermore, the region of uncertainty associated with the new estimate is smaller than either of the two regions of uncertainty associated with conservation or neighborhood information.

Performance at motion boundaries. In quantifying neighborhood information, I have assumed so far that the neighborhood velocity distribution forms a single cluster in *uv*

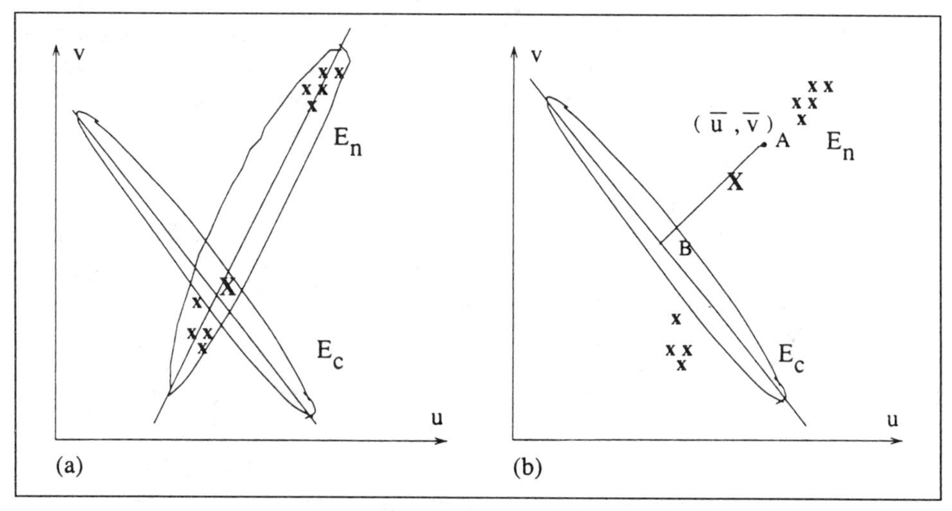

Figure 5.8. Performance at motion boundaries: (a) procedure using neighborhood information, (b) procedure using conventional smoothing.

space. Obviously, this assumption does not hold true at motion boundaries, surfaces slanting away from the viewpoint, rotating surfaces, and so on. In the following discussion, I analyze the performance of the framework at motion boundaries—the other scenarios are similar. Specifically, I show that the procedure discussed above for using neighborhood information is still justified. In the absence of texture, it performs better at the discontinuities of the flow field than the conventional smoothing-based procedure [34, 80]. (In this context, *texture* implies an intensity variation whose scale is smaller than the size of the search window.)

For this discussion, recall that each of the two estimates U_{cc} and $\overline{U}$ maps to a point in uv space. Similarly, each of the two covariance matrices S_{cc} and S_n maps to an ellipse that has its center at the respective estimate and that has its major and minor axes equal to the eigenvalues of the covariance matrix. Therefore, each iteration amounts to finding a point in uv space that has the minimum weighted sum of squared perpendicular distances from the axes of the two ellipses—the eigenvalues serving as weights.

Figure 5.8a shows the behavior of this procedure in the vicinity of a motion boundary. For the conservation ellipse E_{cc}, only the major axis is shown because the minor axis will be very small in this region. In other words, all that the conservation information says (with high confidence) about the velocity of the central pixel is that it lies somewhere along the major axis of the ellipse E_{cc}. Velocities of neighboring points are also plotted from the previous iteration. Assuming that there is no texture in the vicinity of the boundary, that the intensity is smoothly varying (i.e., conservation information is reliable), and that the boundary corresponds to a step discontinuity in the flow field, the velocities of neighboring points form two clusters in uv space. As a result, the minor axis of the neighborhood ellipse E_n will be very small. In other words, all that the neighborhood information says (with high confidence) about the velocity of the central pixel is that it lies somewhere along the major axis of the ellipse E_n. Since the correct velocity will lie in one of the two clusters, this opinion of neighborhood is correct. In other words, the iterative update procedure developed for the nonboundary pixels is justified even for pixels that lie on a motion boundary.

To demonstrate that this method outperforms conventional smoothing at discontinuities, Figure 5.8b shows the result of conventional smoothing [34, 80]. As discussed in Chapter 3, a smoothing procedure such as that of Anandan [80] or Horn and Schunck [34] will place the updated velocity on the line $\overline{AB}$ in uv space that is perpendicular to the major axis of E_{cc} and that passes through the point $(\overline{u}, \overline{v})$ corresponding to the average velocity in the local neighborhood. This is clearly inappropriate because we know with high confidence that the updated velocity lies on the major axis of E_n. Also, as discussed in Chapter 3, this leads to blurring of the flow field at the discontinuity.

On the other hand, the propagation procedure discussed in this section places the updated velocity (approximately) at the intersection of the two major axes. This is justified because there is a high confidence associated with each of the two major axes. Further, the updated velocity will be closer to the cluster with which the conservation information at the pixel is most consistent.

Some implementation issues. The implementation issues that need to be addressed are

- selection of the neighborhood size to compute neighborhood information,
- assignment of weights to the points in uv space that correspond to velocities of the pixels in the neighborhood selected above, and
- inversion of matrices for the iterative procedure given by Equation 5.15.

Selecting too large a neighborhood has a blurring effect on the recovered flow field, so a small neighborhood is preferable. I use a 3×3 (eight-connected) neighborhood. In other words, the parameter w is set to 1. Weights are assigned to points in uv space according to the distance of the corresponding pixel from the central pixel in the neighborhood—the smaller the distance, the larger the weight. I use a 3×3 Gaussian mask for this purpose. That is, the points that correspond to the horizontal and vertical neighbors of the central pixel get a weight of 0.1250, and those corresponding to the diagonal neighbors get a weight of 0.0625. Conservation information is computed only once at the onset, and neighborhood information is computed once for each iteration.

Finally, the iterative solution given by Equation 5.15 breaks down when any one of the two covariance matrices is singular, and hence is noninvertible. Even when a matrix is not singular, one or more of its eigenvalues may be so small that matrix inversion may be ill-conditioned. Such a situation arises in the vicinity of very sharp and high-contrast edges and corners, exactly the regions that are very rich in information content. Hence, rather than compute the matrix inverse directly, I compute it using singular-value decomposition [149, 150]. This allows special handling of cases where one or more eigenvalues are very small or equal to zero.

Let the 2×2 matrix under consideration be denoted by S. It can be written as the product of the following three matrices:

- a 2×2 orthogonal matrix S_1,
- a 2×2 diagonal matrix S_2 with only positive or zero elements, and
- the transposition of a 2×2 orthogonal matrix S_3.

That is,

$$S = S_1 \times S_2 \times S_3^T \tag{5.17}$$

Equation 5.17 can always be decomposed for a 2×2 covariance matrix using standard algorithms [150]. Let the diagonal elements of S_2 be s_{11} and s_{22}, respectively. Then, S^{-1} can be written as

$$S^{-1} = S_3 \times \begin{pmatrix} \frac{1}{s_{11}} & 0 \\ 0 & \frac{1}{s_{22}} \end{pmatrix} \times S_1^T \tag{5.18}$$

The case that requires special handling is where s_{11} or s_{22} is zero (or both are zero). Typically (in solving a system of linear equations, for example), if s_{ii} is zero, $1/s_{ii}$ in Equation 5.18 is replaced by zero. In our case, a small s_{ii} corresponds to a small eigenvalue, and hence a high confidence in the motion estimate (in the direction of the corresponding eigenvector). Hence, the corresponding $1/s_{ii}$ should be replaced by a high value rather than zero. It should, however, be prevented from reaching infinity.

In all the implementations shown in this book, the upper limit on the value of any $1/s_{ii}$ is 10. That is, whenever the value of $1/s_{ii}$ is greater than 10, it is replaced by 10. This implies that an eigenvalue of less than 0.1 is not allowed. Further, the assumption of unimodality ensures that eigenvalues much larger than the window sizes do not exist. Of the various images used in the experiments reported in this book, none of the eigenvalues exceeds 6.23. In essence, the heuristic for singular-value decomposition is based on the actual magnitude of the eigenvalues. The relative sizes of the two eigenvalues is an important attribute of the system (for instance, if the two eigenvalues are almost equal, there is a high uncertainty in the directions of the eigenvectors). This information can be used to come up with better heuristics.

Figure 5.9 shows the results of application of this procedure to the running example. Again, the intent is only to illustrate the principle, with the detailed experimental evaluation deferred until Chapter 6. Conservation information is obtained using the correlation-based approach as discussed earlier in this chapter (see Figure 5.5). Figures 5.9a and 5.9b show respectively one of the frames of the sequence and the correlation-based initial estimates. The iterative update procedure is applied using the various

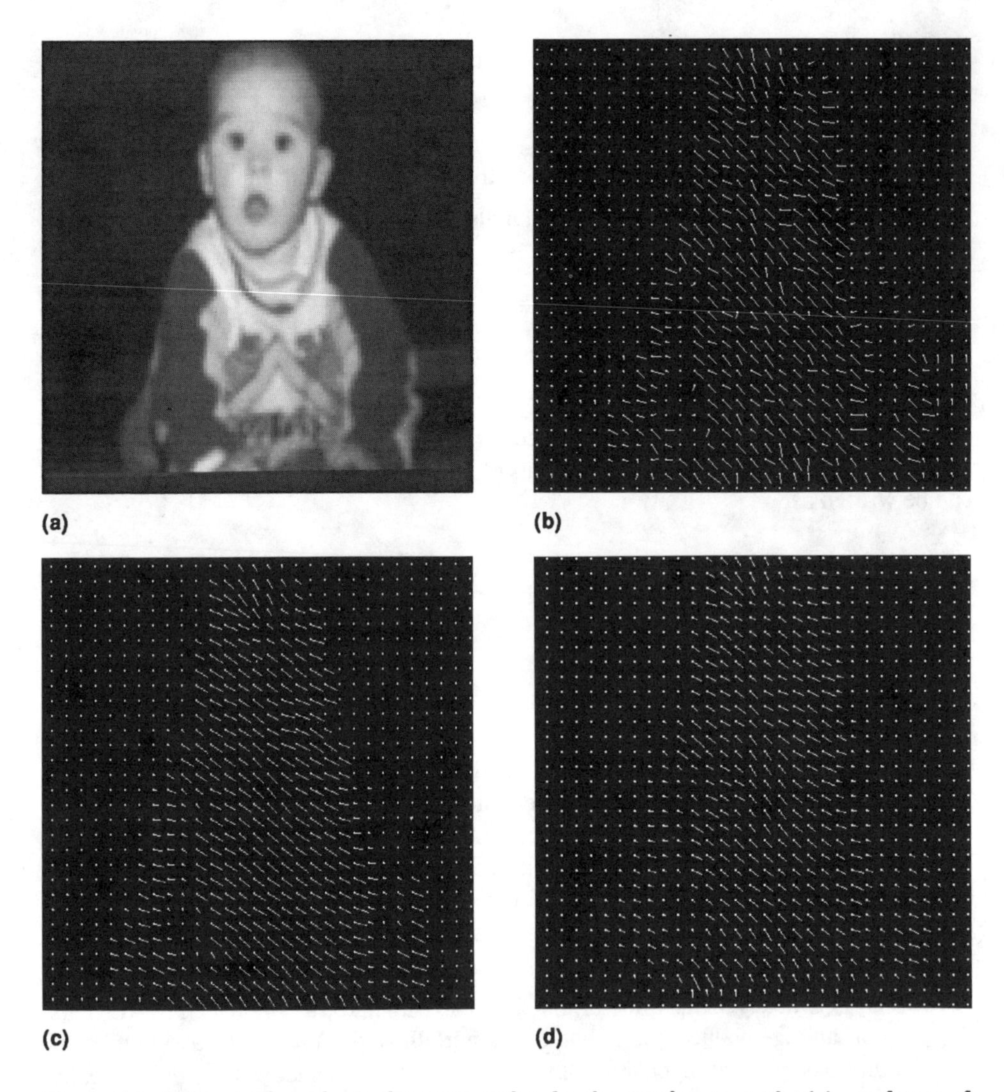

Figure 5.9. An illustration of velocity propagation for the running example: (a) one frame of the original image sequence, (b) correlation-based initial estimates, (c) result after 15 iterations of the new update procedure, (d) result after 15 iterations of conventional smoothing.

parameter values discussed above. Figure 5.9c shows the result after 15 iterations. For comparison, Figure 5.9d shows the result after 15 iterations of conventional smoothing [34, 80] with the same parameters. The motion boundaries are much more crisp in Figure 5.9c. Notable regions are the baby's left ear (right in the picture) and the right arm (left in the picture).

The cost of each iteration is less in conventional smoothing by about a factor of two, counting the number of multiplications. Also, conventional smoothing converges in about two to four fewer iterations in all the experiments reported here. The primary issue of comparison, however, is the behavior at motion discontinuities.

Relationship with major current approaches. A distinct advantage of the scheme for incorporating neighborhood information discussed above is that it provides a simple mechanism to prevent blurring of the flow field at motion boundaries. Some of the current approaches for velocity propagation do have mechanisms for this purpose. Notable among them are those of Hutchinson, Koch, and Mead [141], Aisbett [140], Schunck [76], Nagel [73, 74], and Waxman and Wohn [102]. Also, Anandan's scheme [80] has a provision for incorporating motion boundaries, although it has not been implemented. I briefly compare the method shown in this section with these existing techniques.

The recent technique suggested by Hutchinson, Koch, and Mead [141] draws on the theory of Markov random fields [151] and applies the powerful idea of line processes as a supplement to Horn and Schunck's method to explicitly prevent smoothing across motion boundaries. This technique has shown remarkable promise on both synthetic and real imagery. However, it uses a priori knowledge about the possible locations of motion boundaries. Regions with either high-intensity gradients or high-velocity gradients are marked as possible motion boundaries and line processes are started at these locations. The iterative update procedure generates a flow field with appropriate motion boundaries. If an intensity discontinuity does not correspond to a motion discontinuity, the corresponding line process eventually dies.

My procedure is different from that of Hutchinson, Koch, and Mead because a boundary evolves during the propagation process only if there is a motion discontinuity suggested by conservation information (in the form of multiple clusters in the neighborhood velocity distribution). Also, this method is computationally much less expensive.

Aisbett's approach [140] is essentially smoothing based and is similar to that of Horn and Schunck. However, the smoothness term in the minimization constraint is weighted by the underlying intensity gradient so that the regions of high-intensity gradient are allowed to have large values of velocity gradient. Indirectly, this method allows discontinuities to emerge in the smoothed flow field in the regions of strong intensity gradients. This approach has an underlying shortcoming in that it assumes that all intensity edges correspond to motion boundaries and that motion boundaries occur only at intensity edges. Obviously, this assumption is not always true. Similarly, Schunck's approach [76] for discontinuous flow fields introduces the novel idea of constraint-line clustering. However, this approach also relies, in part, on identification of motion boundaries as discontinuities in the underlying intensity function.

Nagel's approach introduces the idea of an oriented smoothness constraint, which was described in Chapter 3. In essence, the regions where the intensity function has a high gradient and/or a high curvature in a particular direction are allowed to violate the smoothness criterion in that direction. Thus, the flow field is not blurred in the regions of high gradient or high curvature. However, this approach also assumes that all such regions correspond to motion boundaries. The assumption is not always true.

I discussed Waxman and Wohn's approach [102] in Chapter 3. They suggest that motion boundaries can be detected during the minimization process by thresholding the magnitude of error residual. Their approach has shown promising results on synthetic imagery but its performance with real imagery is yet to be established.

Algorithms

Here I describe two algorithms based on the new framework. Both use three images as their input. The first algorithm uses the input imagery at its original resolution and is applicable when velocity is known to be very small. It recovers conservation information only once at the onset (steps 1 through 3) and neighborhood information once for each iteration (steps 4 through 6). The second algorithm uses a multiresolution strategy and is applicable when velocity can range from very small to very large.

Algorithm 1

1. Perform band-pass filtering on each image. That is, compute the highest resolution image in Burt's Laplacian pyramid.
2. Form a $(2n + 1) \times (2n + 1)$ correlation window around the pixel under consideration in the central image. Also, form a $(2N + 1) \times (2N + 1)$ search window around the corresponding location in the other two images. Compute the error distributions over the two search windows using Equation 5.1 and transform them to the corresponding response distributions $\mathcal{R}_c^{-1}$ and $\mathcal{R}_c^{+1}$, respectively, using Equation 5.2. Finally, use $\mathcal{R}_c^{-1}$ and $\mathcal{R}_c^{+1}$ to compute the resultant response distribution $\mathcal{R}_c$.
3. Compute the estimate U_{cc} and the covariance matrix S_{cc} from the response distribution using Equations 5.3 and 5.4, respectively.
4. Form a $(2w + 1) \times (2w + 1)$ window around the pixel under consideration. Denote each pixel by a distinct index i, where $1 \leq i \leq (2w + 1)^2$. Denote the current estimate of the velocity of the ith pixel by (u_i, v_i). (For the first iteration, the velocity U_{cc} computed in step 3 can be used as the current estimate.) Assign weights $\mathcal{R}_n(u_i, v_i)$ to these velocities. Compute the mean $\overline{U}$ and the covariance matrix S_n using Equations 5.6 and 5.7, respectively.
5. Update the velocity at the pixel under consideration using Equation 5.15. Use the techniques suggested in the earlier section on neighborhood information implementation issues to invert various matrices.
6. Repeat steps 4 and 5 until the change in velocity over two successive iterations is less than a threshold.
7. Compute the confidence measures associated with the final estimate of velocity as the eigenvalues of the matrix given by $S_{cc}^{-1} + S_n^{-1}$. These confidence measures are

associated with the directions of maximum and minimum confidence—that is, along the eigenvectors.

For completeness, this algorithm must have a mechanism to deal with a pixel that lies on the image boundary. I simply replicate the boundary pixel X times (in effect, increasing the image size by X pixels beyond each boundary), where X is the largest of the correlation window radius n, the search window radius N, and the neighborhood radius w. This enables the algorithm to accommodate all three windows around the boundary pixels.

Algorithm 2. Algorithm 1 is applicable only if image velocity is small compared with the size of the search window. A hierarchical scheme is required to estimate large velocities. Here, the large velocities are estimated approximately at a coarse resolution and refined at finer resolutions. Small velocities, on the other hand, are estimated at fine resolutions directly. Several such schemes have been proposed in the past research [68, 80]. I use a variant of Anandan's scheme [80] in the following algorithm. Issues related to hierarchical implementation need further research and are discussed in Chapter 10.

1. Construct a band-pass (Burt's Laplacian) pyramid from each image in the sequence, up to a resolution where the maximum possible image displacement over two successive frames is less than the radius of the search window. Label the various levels of the pyramid *level 0*, *level 1*, and so on; level 0 is the finest resolution.
2. Run Algorithm 1 at the coarsest level. This gives a coarse estimate of velocity at each pixel.
3. Project the velocity estimate at each pixel to the four "children" pixels at the next finer resolution. The projected magnitude of each component of velocity is *twice* the magnitude at the coarser resolution. In essence, this procedure gives an approximate estimate of the flow field at the current level of resolution.
4. Refine the approximate estimate obtained above as follows. Let the pixel under consideration be located at (x, y) and the current estimate of its velocity be given by (u, v). Run Algorithm 1 at the current resolution with the following change in step 2 (of Algorithm 1). Instead of forming the $(2N + 1) \times (2N + 1)$ search window around the location (x, y), form it around the location $(x + u, y + v)$, because the search for the matching pixel does not have to be "blind" anymore. A rough estimate of velocity is available from the coarser resolutions and it guides the search.
5. Repeat steps 3 and 4 until the velocity estimate at the finest resolution is obtained.

Conclusion

In this chapter, I have shown a new framework for recovering optic flow from time-varying imagery. This framework recognizes the fact that the velocity information available in small spatiotemporal neighborhoods in the imagery is not exact; there is uncertainty associated with it. It classifies the available information into two categories—conservation information and neighborhood information—and models each using techniques common in estimation theory. It recovers the optic-flow field by performing an MSE-optimal combination of the two types of information. Some of the distinctive features of the framework are summarized below:

- It quantifies the velocity information contained in each of the two local sources—conservation and neighborhood—by an estimate and a covariance matrix. A similar approach has been used before [80] for conservation information. However, as far as neighborhood information is concerned, this approach is novel. In essence, the current formulation accounts for the spread (in velocity space) of neighborhood velocities, in addition to their average, as has been used in earlier formulations [34, 71].

- It formulates the problem of estimating optic flow as that of performing a statistical combination of velocity estimates obtained from the two sources on the basis of their covariance matrices. The solution to this problem is iterative and amounts to propagating velocity information from regions of low uncertainty to regions of high uncertainty.

- Because of the statistical nature of the procedure used to represent and propagate velocity, there is an explicit notion of confidence measures associated with the velocity estimate at each pixel, both before and after propagation. Prepropagation confidence measures have been used before [80], but postpropagation confidence measures are novel. (Szeliski [152] also uses the notion of a posteriori confidence measures on the basis of a Bayesian model. However, these confidences are not a direct consequence of velocity propagation.). Experiments in the next chapter will reveal that the iterative propagation procedure used in this framework does actually enhance the confidence during each iteration. The postpropagation confidence measure reflects the reliability of the final estimate of optic flow and it can be a valuable input to a high-level interpretation system that uses optic flow to recover 3D information. In Chapter 7, I show that the postpropagation confidence measures are used directly by a Kalman-filtering-based technique to recover depth information from optic flow.

- The propagation procedure does a much better job than the classic smoothing-based propagation procedures [34] of preserving the step discontinuities in the flow field, especially in the absence of texture in the vicinity of such discontinuities. I have demonstrated this with the baby sequence of the running example. I verify this claim more thoroughly in the next chapter.

This framework does have its pitfalls:

- The procedure to recover the initial estimate of velocity (i.e., conservation information) is based on the assumption of conservation of some image property (band-pass filtered intensity, in the current implementation) over time. In reality, there are inevitable deviations from the conservation assumptions (because of photometric changes, perspective distortions, rotation, and so on). This framework does not address the issue of how to account for these deviations.

- The procedure to recover conservation information is based on the assumption of unimodality of response distribution. This assumption is likely to be violated in the presence of texture or occluding or transparent moving objects, leading to an incorrect estimate. However, the incorrect estimate has a low confidence associated with it. If there is a high-confidence region in the vicinity and the velocity estimate in this region is correct, the incorrect estimate may improve during velocity propagation. If, however, there is no such region nearby, the incorrect estimate stays incorrect even after velocity propagation. I have shown that using more than two frames serves as a useful heuristic in these regions. However, to deal with multimodality more formally, some mechanism must be developed to identify the correct peak in the response distribution and use it to recover the initial estimate of velocity.
- The procedure to assign a covariance matrix (and thus, confidence measures) to the velocity estimate is based on the assumption of additive, zero-mean, and independent errors. I have not done a detailed investigation of either the validity of these assumptions or the effects of violating them on the reliability of the confidence measures. An understanding of these effects may help in developing a robust procedure to interpret the confidence measures.
- The underlying estimation technique used in this framework is that of least squares. It is well known that any scheme based on least squares does not perform well in the presence of outliers.

There are several ways in which this framework can be extended and improved. I describe some of them in the last chapter of this book. The foremost among the extensions is the issue of incremental estimation of optic flow and scene depth over time. I address this issue in the two appendixes that follow the last chapter. In the next chapter, I use the algorithms developed in this chapter to recover optic flow from a variety of image sequences. Later, in Chapter 8, I show that this framework acts as a platform to unify correlation-based, gradient-based, and spatiotemporal energy-based approaches for recovering conservation information and various smoothing-based approaches for performing velocity propagation.

Chapter 6

Experiments on Estimation of Optic Flow

This chapter is devoted to an experimental evaluation of the framework developed in the last two chapters. The experiments described here can be divided into two categories: quantitative and qualitative. I give a detailed description of the objectives, methodology, and results of each category of experiments in the following two sections. In each category, I compute flow fields from image sequences corresponding to a variety of scenes. The criteria for selecting the values of various parameters used in these experiments have already been discussed in the previous chapter and are not repeated here.

Quantitative experiments

The general objective of quantitative experiments is to judge the quantitative correctness of the flow fields and to analyze their behavior with respect to various assumptions and parameters used in the framework. To accomplish this, the "ground-truth" flow field must be known. Typically, it is possible to know (or compute) the ground-truth flow field only if the motion is synthetically generated, for example, by warping a given image in some known fashion, or the camera motion and the depth of each point in the scene are exactly known. Both of these scenarios are considered in the experiments that follow. I refer to them by the scene imaged in them: the Netherlands aerial imagery experiment, the poster experiment, and the fused-image experiment.

In the first two experiments, the imagery is selected so the flow fields do not have any discontinuities, simply because it is very difficult to come up with the ground-truth flow field in the presence of discontinuities. Thus, these experiments do not verify that motion discontinuities are preserved. The third experiment uses imagery for which the underlying flow field does have a motion discontinuity, although it is synthetically generated. The issue of motion discontinuities is addressed further in the qualitative experiments in the next section.

The Netherlands aerial imagery experiment. In this experiment, I use real images that have been displaced with respect to each other manually. In other words, the motion is synthetic and generated directly in the image plane. The correct optic flow is known

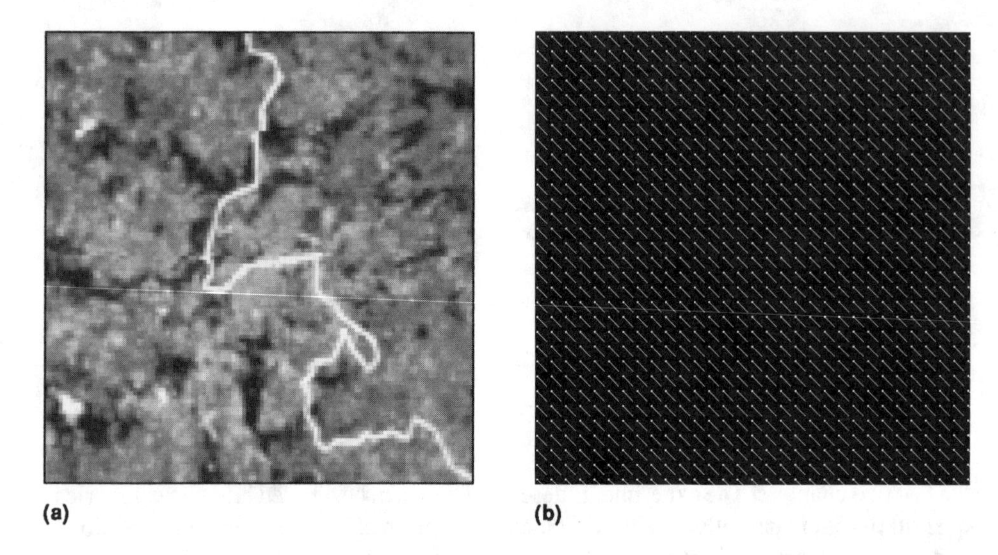

(a) (b)

Figure 6.1. The Netherlands aerial imagery experiment: (a) central frame of the image sequence, (b) correct flow.

exactly. Specifically, three subimages, 124 pixels wide and 120 pixels high, are extracted from a 256×256 aerial image of The Netherlands. These subimages are displaced from each other by one pixel in the horizontal direction and one pixel in the vertical direction. Figure 6.1a shows the central frame of this three-image sequence. The correct velocity at each pixel is composed of a horizontal and a vertical component of one pixel per frame and is shown in Figure 6.1b. Since the velocity is known to be small, Algorithm 1 is used to estimate optic flow.

To compute optic flow, the images are used directly (without any noise added). The search window size is set to 5×5 and the neighborhood window size for velocity propagation is set to 3×3. Two cases are studied, with correlation window sizes set to 5×5 and 3×3, respectively. Since there is no motion boundary, the optic-flow estimates are not very sensitive to an increase in the neighborhood window size to, say, 5×5. The corresponding results, therefore, are not reported. For both values of the correlation window size, the percentages of pixels that have the magnitude of the vector error in velocity within 5 percent (of the true value), within 10 percent, and within 25 percent are determined. The iterative update procedure is terminated when the value of each component of velocity, rounded to the second decimal digit, does not change at any pixel. For this imagery, about 15 iterations are found sufficient. Table 6.1 shows the results.

For both the 3×3 and 5×5 correlation windows, Table 6.2 shows the mean and the standard deviation of each component of velocity (after propagation) computed over a sample comprising all the pixels in the image. As expected, the larger size correlation window (5×5) gives more accurate results, although reasonable results are obtained with the 3×3 correlation window also—especially after velocity propagation. Figure 6.2 graphs the actual distribution of the horizontal and vertical components of velocity

Table 6.1. Error statistics for the Netherlands aerial imagery experiment. The two rows correspond to two different sizes of the correlation window. For each row, the first and second columns indicate the percentages of total pixels for which the magnitude of the vector error in velocity is less than 5 percent before and after velocity propagation, respectively. The third and fourth columns give the corresponding percentages of pixels with error less than 10 percent. Finally, the fifth and sixth columns give the corresponding percentages of pixels with error less than 25 percent. The rows and columns closest to the image border are not used in the computation of these statistics.

Window Sizes	Percentage of pixels with vector error less than 5 percent		Percentage of pixels with vector error less than 10 percent		Percentage of pixels with vector error less than 25 percent	
	Without Propagation	With Propagation	Without Propagation	With Propagation	Without Propagation	With Propagation
5 × 5 Search 3 × 3 Correlation	58.6	64.1	72.4	82.4	77.1	90.6
5 × 5 Search 5 × 5 Correlation	62.1	69.5	77.2	88.2	81.0	94.7

Table 6.2. Mean and standard deviation of optic-flow components for the Netherlands aerial imagery experiment. Postpropagation velocity is used in the computations.

Window Sizes	Horizontal Component		Vertical Component	
	Mean	Standard Deviation	Mean	Standard Deviation
3 × 3 Correlation	0.82	0.21	0.85	0.23
5 × 5 Correlation	1.09	0.16	1.06	0.17

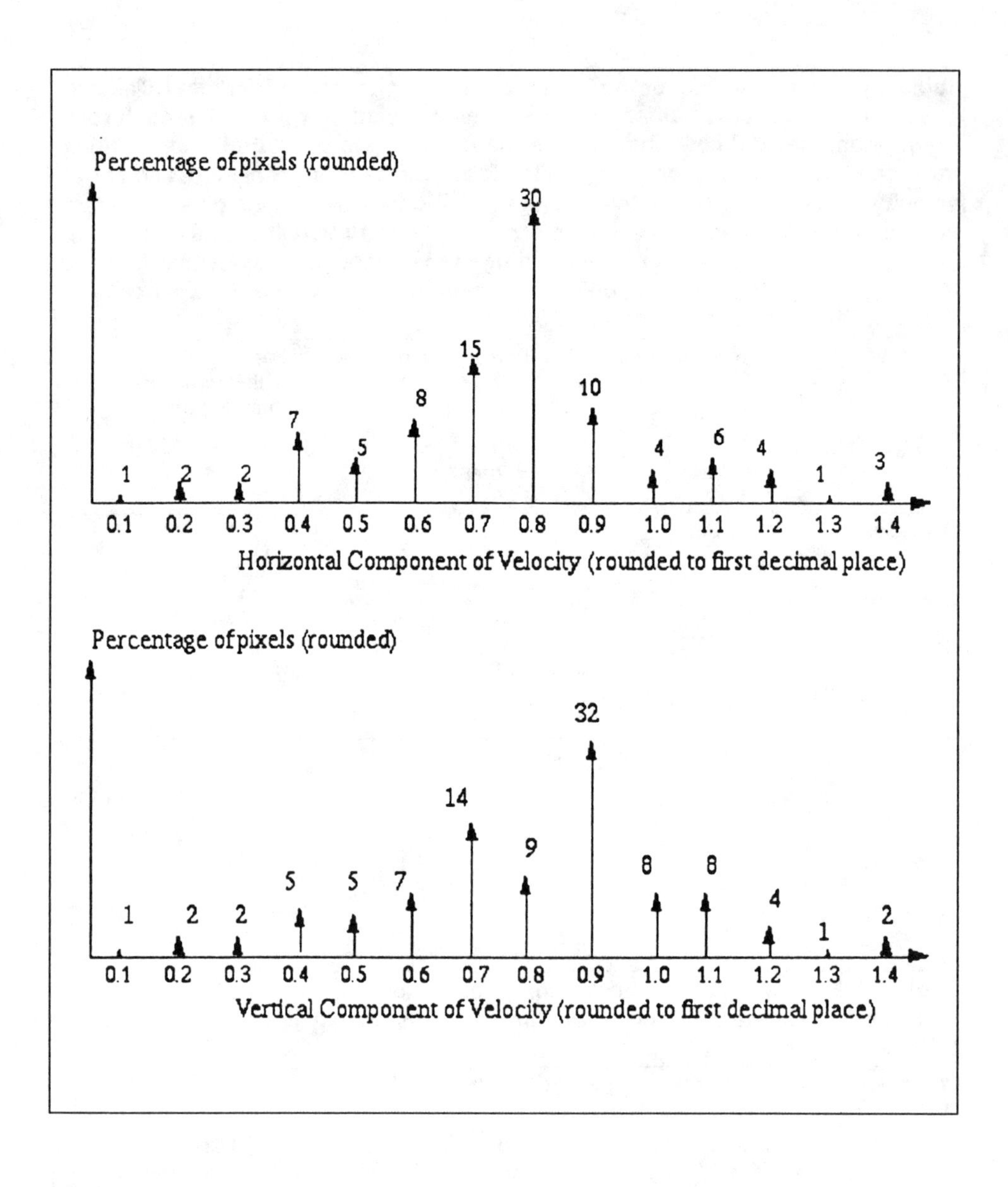

Figure 6.2. The Netherlands aerial imagery experiment: distribution of horizontal and vertical components of velocity with respect to the number of pixels. Postpropagation velocity obtained with a 3×3 correlation window is used in the computations. Because of rounding effects, all the percentages may not sum up to 100.

Table 6.3. Confidence-based error statistics for the Netherlands aerial imagery experiment. The three rows correspond to pixels where each confidence measure is within 25, 50, and 100 percent, respectively, of the highest confidence observed in the image. For each row, the first, second, and the third columns indicate the percentages of pixels under consideration for which the magnitudes of the vector error in the postpropagation velocity are less than 5, 10, and 25 percent, respectively. The rows and columns closest to the image border are not used in the computation of these statistics.

Confidence Regions (in percent)	Percentage of pixels with vector error less than 5 percent	Percentage of pixels with vector error less than 10 percent	Percentage of pixels with vector error less than 25 percent
Top 25	97.4	99.2	100.0
Top 50	75.5	81.2	93.5
Top 100	64.1	77.1	90.6

obtained with the 3×3 correlation window. For the rest of the computations in this experiment, I used a 3×3 correlation window.

Recall that the framework gives confidence measures in addition to the estimates. It is interesting to observe the (postpropagation) estimates for only those pixels where both confidence measures are, say, within 25 percent or 50 percent (instead of the 100 percent used so far) of the highest confidence observed in the image. Table 6.3 gives the percentages of pixels that have their estimates (obtained using the 3×3 correlation window) within 5, 10, and 25 percent of the true value, for each of these confidence regions.

Figures 6.3 through 6.7 show various flow fields and confidence measures. Figure 6.3a shows one frame of the original sequence. Figures 6.3b and 6.3c show the two confidence measures associated with conservation information (i.e., the initial estimate of velocity) at each point in the visual field. These confidence measures are the inverses of the small and the large eigenvalues, respectively, of the covariance matrix S_{cc}. Figure 6.3d shows the initial estimate of the flow field (i.e., the velocity U_{cc}). Figure 6.4 shows the flow field after iterative velocity propagation (15 iterations). Figure 6.5 shows the actual magnitude of the horizontal component of the flow field (scaled up by a factor of 10 and rounded to the nearest integer). The vertical component is similar and is not shown here. For comparison, Figure 6.6 shows the flow field after 15 iterations of conventional smoothing [34, 80] with the smoothing factor α set to 0.5. Figures 6.7a and 6.7b show the two confidence measures after propagation. It is apparent that the confidence has propagated outward from the prepropagation high-confidence regions.

Next, the effects of adding noise to the imagery are studied. Random noise with a zero-mean Gaussian distribution is added to the first and third images. Three tests are

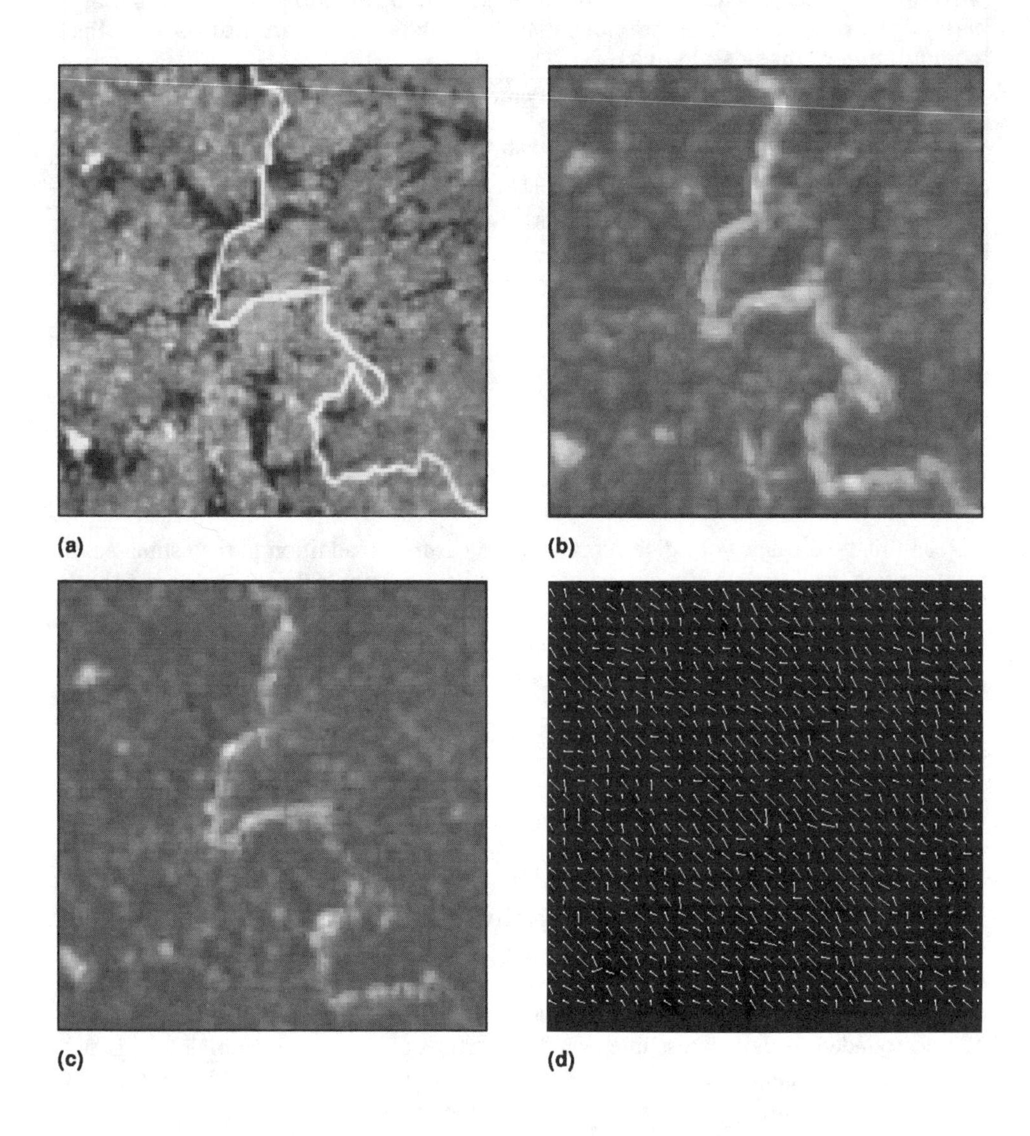

Figure 6.3. The Netherlands aerial imagery experiment: (a) central frame of the image sequence, (b, c) confidence measures associated with conservation information, that is, the reciprocals of the eigenvalues of the covariance matrix S_{cc}, (d) initial estimate of velocity U_{cc}.

Figure 6.4. The Netherlands aerial imagery experiment: flow field after velocity propagation (15 iterations).

1	2	3	2	4	11	4	8	8	10	13	6	9	3	2	2	2	6	3	2	2	1
3	6	7	6	7	6	5	8	8	8	9	13	10	6	4	5	7	7	7	7	6	3
8	8	7	7	7	5	5	7	7	7	9	14	10	5	4	6	7	8	8	8	6	4
9	10	10	9	9	7	6	7	7	5	15	13	8	6	7	8	9	8	8	8	6	5
6	11	12	8	8	8	9	8	6	6	11	11	8	8	9	8	8	8	8	7	8	7
2	6	8	8	7	8	9	7	6	8	14	11	9	10	10	7	6	8	8	6	6	4
2	5	7	9	8	8	8	6	6	10	15	12	9	9	8	7	7	9	8	6	5	3
4	6	6	9	10	8	6	6	10	13	13	11	8	8	8	7	7	8	9	8	6	2
7	8	6	8	8	8	7	9	13	12	11	9	7	10	11	9	7	7	7	8	7	3
8	8	7	6	6	6	8	13	15	10	8	7	7	11	11	10	7	6	6	8	8	4
9	7	8	6	6	6	8	12	15	11	11	11	12	13	11	9	7	6	7	8	8	3
2	5	8	7	8	8	10	12	14	13	10	8	9	14	11	9	8	8	8	7	6	7
2	6	7	8	9	8	10	11	12	12	8	6	7	10	10	9	8	7	8	7	7	7
6	7	7	8	8	7	9	10	10	10	8	8	7	7	9	11	10	7	7	7	7	3
5	8	8	8	7	6	7	8	9	9	9	9	7	5	9	12	12	9	7	7	5	3
2	8	8	8	7	8	8	10	9	9	11	10	10	13	13	11	12	9	6	6	3	2
4	8	8	8	9	10	10	10	9	10	11	10	8	14	11	9	9	7	6	5	4	5
7	12	10	8	9	9	10	9	9	11	13	9	1	14	10	7	6	5	7	7	6	5
9	12	7	7	8	7	7	9	11	11	13	9	8	12	11	8	8	7	11	8	6	2
10	8	7	7	7	6	6	9	9	8	9	9	7	9	10	9	9	8	8	7	9	10
2	1	1	2	2	1	1	13	3	2	8	14	2	3	6	2	2	3	2	3	11	13

Figure 6.5. The Netherlands aerial imagery experiment: magnitude of the horizontal component of the flow field after velocity propagation (multiplied by a scale factor of 10 and rounded to the nearest integer).

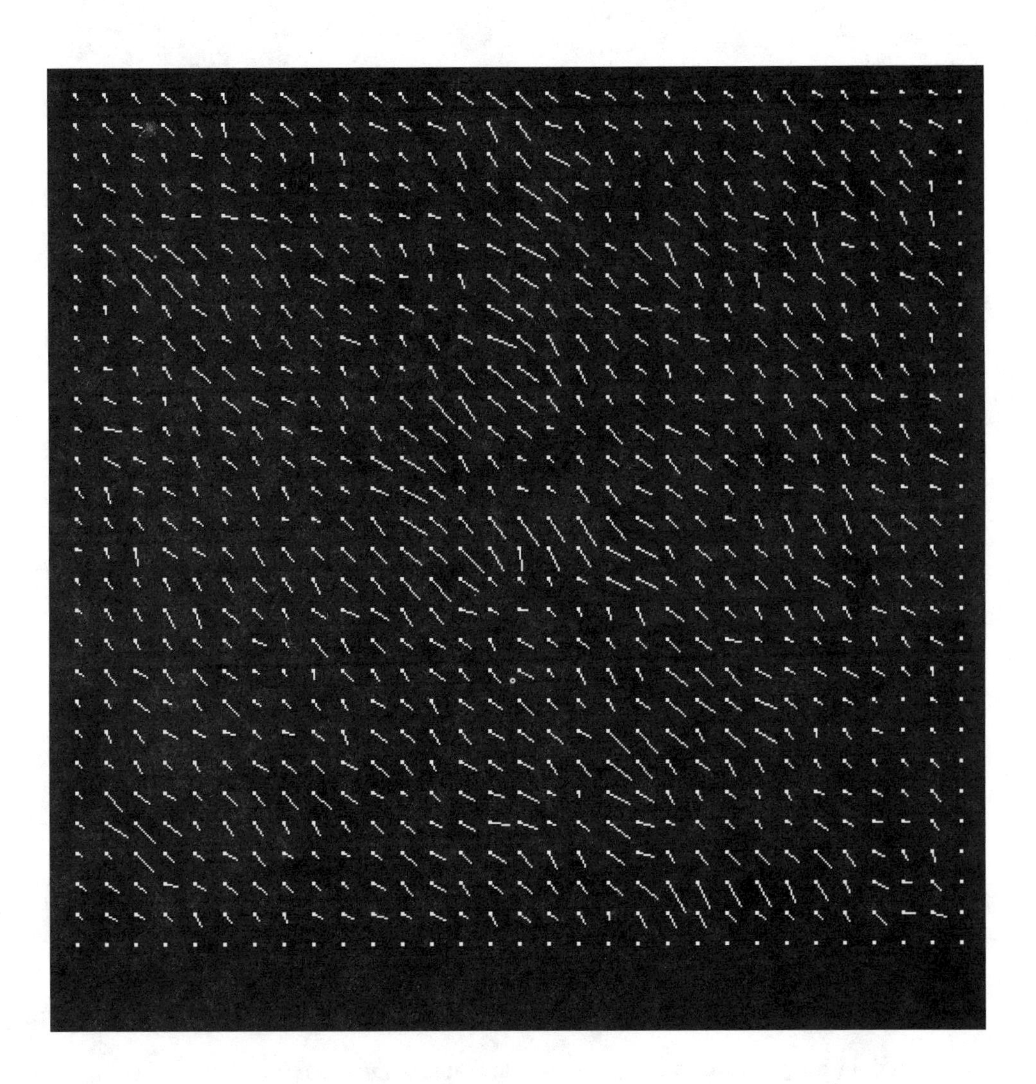

Figure 6.6. The Netherlands aerial imagery experiment: flow field after 15 iterations of conventional smoothing (the algorithm converged after 10 iterations).

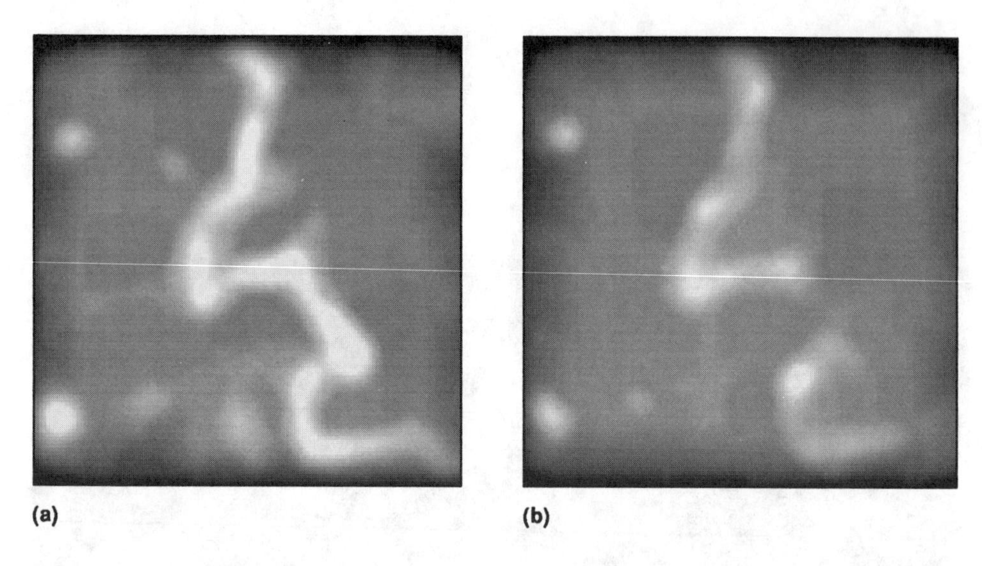

(a) (b)

Figure 6.7. The Netherlands aerial imagery experiment: confidence measures associated with the flow field after velocity propagation.

Table 6.4. Error statistics for various amounts of noise in the Netherlands aerial imagery experiment.

Noise (in percent)	Percentage of pixels with vector error less than 5 percent		Percentage of pixels with vector error less than 10 percent		Percentage of pixels with vector error less than 25 percent	
	Without Propagation	With Propagation	Without Propagation	With Propagation	Without Propagation	With Propagation
5	52.4	61.2	60.9	78.1	72.6	87.0
10	31.3	56.3	46.9	71.3	56.0	82.3
25	10.5	38.1	20.3	45.9	41.2	75.1

conducted with the standard deviation of the Gaussian distribution set to 5, 10, and 25 percent of the average intensity of the image. Table 6.4 indicates the percentages of pixels that have an error less than 5, 10, and 25 percent in each component of velocity, before and after propagation. It is apparent that the algorithm is quite robust to small amounts of image noise, especially on the order of 5 percent. The effects of salt-and-pepper noise are very similar to those of Gaussian noise and are not reported here.

Table 6.5. Mean and standard deviation of optic-flow components for the Netherlands aerial imagery experiment. Postpropagation velocity is used in the computations.

Window Sizes	Horizontal Component		Vertical Component	
	Mean	Standard Deviation	Mean	Standard Deviation
3 × 3 Search	0.72	0.26	0.66	0.28
9 × 9 Search	1.52	0.36	1.61	0.36

The following discussion illustrates the effects of violating the basic assumptions on which the algorithm is based. These assumptions are

1. The velocity to be estimated lies within the search window.
2. Errors in velocity measurement are additive, have zero mean, and are independent.
3. The response distribution is unimodal within the search window.

The algorithm also assumes that conservation error and neighborhood error are independent. As discussed in Chapter 5, this assumption is clearly violated. In this book, I have not pursued the techniques to make these errors independent. I do, however, list some possible strategies in Chapter 10.

The first assumption above can be violated by decreasing the size of the search window. Generally, the effect of violating the first assumption is to corrupt the estimate of velocity (because the mean will be forced to lie within the search window). Decreasing the size of the search window also violates the second assumption by causing the mean error to deviate from zero. (Within the current framework, it is difficult to violate the assumption of additive errors. The assumption of independent errors is almost always violated. However, it does not seem to have a noticeable effect on the estimates.) The violation of the second assumption does not affect the initial estimate of velocity directly. It does, however, corrupt the estimate of the covariance matrix, and hence corrupts the final estimate of velocity. The third assumption can be violated by using a search window larger than the scale of the texture and using only two images to recover conservation information. As discussed in Chapter 5, the effect of violating the assumption of unimodality is to either underestimate or overestimate the velocity, depending on the relative location of various peaks within the search window.

In this experiment, the first and the second assumptions are violated by setting the search window size to 3×3. Similarly, the third assumption is violated by setting the search window size to 9×9 and using only two images to recover conservation information. Table 6.5 depicts the results. A comparison of the corresponding entries of Table 6.5 and Table 6.2 confirms the effects discussed above.

The poster experiment. In the poster experiment, I used a setting where optic flow is caused by real motion in the scene, rather than synthetic motion in the image. To judge the quantitative correctness of the recovered optic flow, the ground-truth optic flow must be known very accurately. Knowing the correct flow field is usually quite difficult in "real-motion" settings, even more so if the magnitude of image velocity is required to be very small (on the order of one or two pixels per frame). Nevertheless, I made an attempt is this experiment.

Specifically, the scene consists of a textured poster rigidly mounted on a precision translation table. A 512×512 camera is mounted on the table as well, but its (translational) motion can be accurately controlled up to one hundredth of an inch. The poster is placed facing the camera and slanted so the optical axis is not perpendicular to the plane of the poster and the distance between the camera and the poster is very small (about 9 inches). Both these arrangements help to make the resulting flow field interesting, even when the camera is undergoing a pure translation. The camera is made to translate so that the image displacement is six pixels where the poster is closest to the camera and three pixels where the poster is the farthest from the camera. The exact amount of camera translation as well as the distance of the lens from the rigid mount is recorded. The camera is then calibrated and its focal length is determined. The "correct" flow field is determined using the projection equations. The images are low-pass filtered and subsampled to get a resolution of 128×128 using Burt's technique [136]. Both components of image velocity at each point are divided by four to get the correct flow field corresponding to the reduced image size. The reduced-size imagery will actually correspond to optic flow that is not exactly equal to the original optic flow reduced in magnitude by a factor of four. This is because of digitization effects and the intensity changes that accompany low-pass filtering and subsampling. Because of the lack of a quantitative characterization of these changes, I do not account for them.

The central image and the correct flow field are shown in Figures 6.8a and 6.8b, respectively. Once again, since the velocity is known to be small, Algorithm 1 is used to estimate optic flow.

For optic-flow computation, the images are used directly (without any noise being added). The search window size is set to 5×5 and the neighborhood window size for velocity propagation is set to 3×3. Two cases are studied, with the correlation window size set to 5×5 and 3×3, respectively. In each case, the percentages of pixels that have the magnitude of the vector error in velocity within 5 percent (of the true value), within 10 percent, and within 25 percent are determined. The iterative update procedure is terminated when the value of each component of velocity, rounded to the second decimal digit, does not change at any pixel. For this imagery, about 10 iterations are sufficient. Table 6.6 shows the results.

The mean and the standard deviation of the error in each component of velocity (after propagation), computed over a sample comprising all the pixels in the image, are shown in Table 6.7, for both the 3×3 and 5×5 correlation windows. As expected, the larger correlation window (5×5) gives more accurate results, although reasonable results are obtained with the 3×3 correlation window also, especially after velocity propagation.

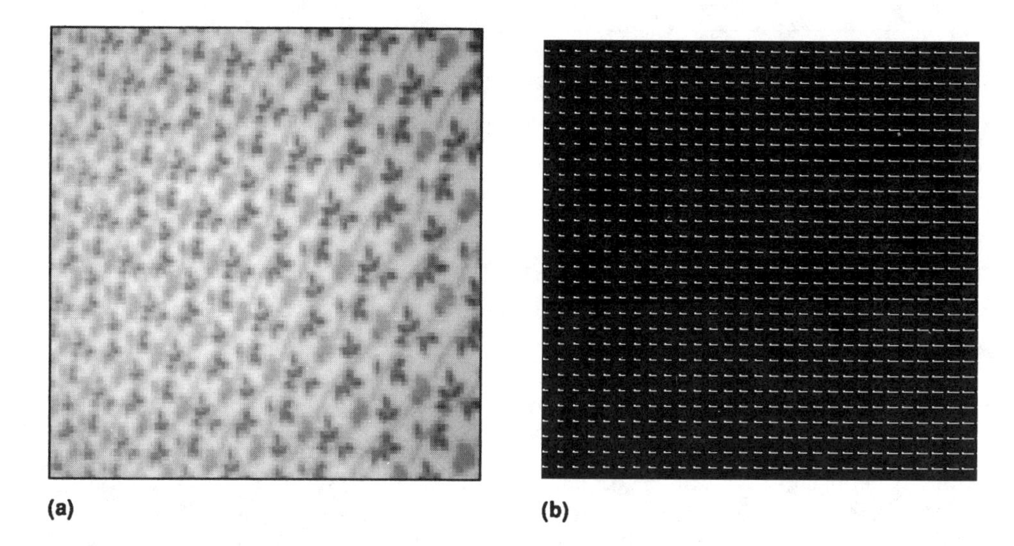

(a) (b)

Figure 6.8. The poster experiment: (a) central frame of the image sequence, (b) correct flow field.

Table 6.6. Error statistics for the poster experiment. The two rows correspond to two different sizes of the correlation window. For each row, the first and second columns indicate the percentages of total pixels for which the magnitude of the vector error in velocity is less than 5 percent before and after velocity propagation, respectively. The third and fourth columns give the corresponding percentages of pixels with error less than 10 percent. Finally, the fifth and sixth columns give the corresponding percentages of pixels with error less than 25 percent. The rows and columns closest to the image border are not used in the computation of these statistics.

Window Sizes	Percentage of pixels with vector error less than 5 percent		Percentage of pixels with vector error less than 10 percent		Percentage of pixels with vector error less than 25 percent	
	Without Propagation	With Propagation	Without Propagation	With Propagation	Without Propagation	With Propagation
5 × 5 Search 3 × 3 Correlation	53.0	56.1	66.4	77.5	71.3	83.1
5 × 5 Search 5 × 5 Correlation	56.2	61.2	68.6	81.6	73.2	86.4

Table 6.7. Mean and standard deviation of the error in optic-flow components for the poster experiment. Postpropagation velocity is used in the computations.

Window Sizes	Error in Horizontal Component		Error in Vertical Component	
	Mean	Standard Deviation	Mean	Standard Deviation
3 × 3 Correlation	0.08	0.06	0.11	0.10
5 × 5 Correlation	0.05	0.03	0.07	0.07

Figure 6.9 shows the actual distribution of the horizontal and vertical components of velocity obtained with the 3 × 3 correlation window. For the rest of the computations in this experiment, I used a 3 × 3 correlation window. Once again, it is interesting to observe the (postpropagation) estimates for only those pixels where both confidence measures are, say, within 25 or 50 percent (instead of the 100 percent used so far) of the highest confidence observed in the image. Table 6.8 gives the percentages of pixels that have their estimate (obtained using the 3 × 3 correlation window) within 5, 10, and 25 percent of the true value, for each of these confidence regions.

Figures 6.10 through 6.14 show various flow fields and confidence measures. Figure 6.10a shows one frame of the original sequence. Figures 6.10b and 6.10c show the two confidence measures associated with conservation information (i.e., the "initial" estimate of velocity) at each point in the visual field. These confidence measures are the inverses of the small and large eigenvalues, respectively, of the covariance matrix S_{cc}.

Table 6.8. Confidence-based error statistics for the poster experiment. The three rows correspond to pixels where each confidence measure is within 25, 50, and 100 percent, respectively, of the highest confidence observed in the image. For each row, the first, second, and third columns indicate the percentages of the pixels under consideration for which the magnitudes of the vector error in the postpropagation velocity are less than 5, 10, and 25 percent, respectively. The rows and columns closest to the image border are not used in the computation of these statistics.

Confidence Regions (in percent)	Percentage of pixels with vector error less than 5 percent	Percentage of pixels with vector error less than 10 percent	Percentage of pixels with vector error less than 25 percent
Top 25	95.1	98.2	100.0
Top 50	70.3	84.1	88.2
Top 100	56.1	77.5	83.1

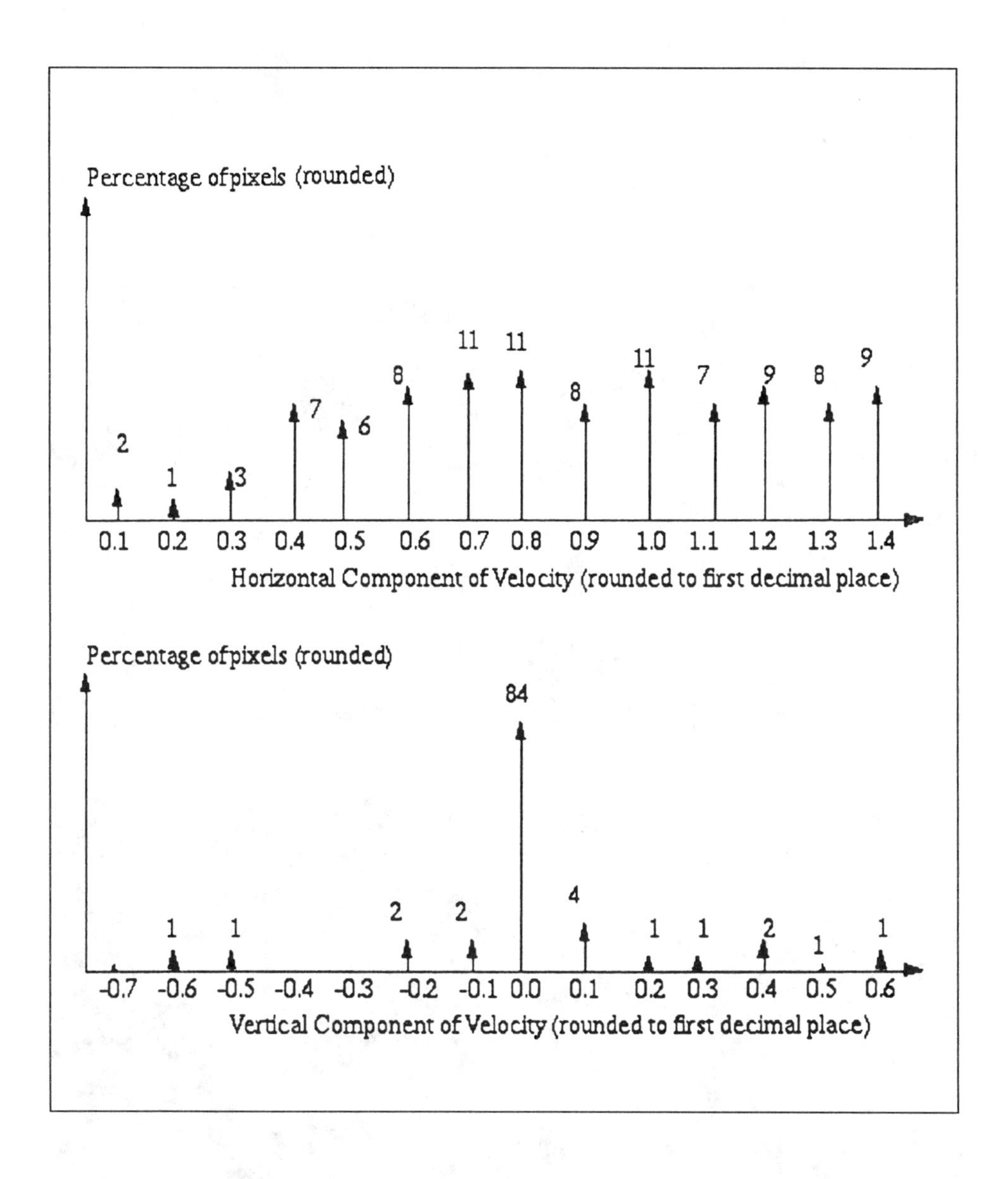

Figure 6.9. The poster experiment: distribution of horizontal and vertical components of velocity with respect to the number of pixels. Postpropagation velocity obtained with a 3 × 3 correlation window is used in the computations. Because of rounding effects, all the percentages may not sum up to 100.

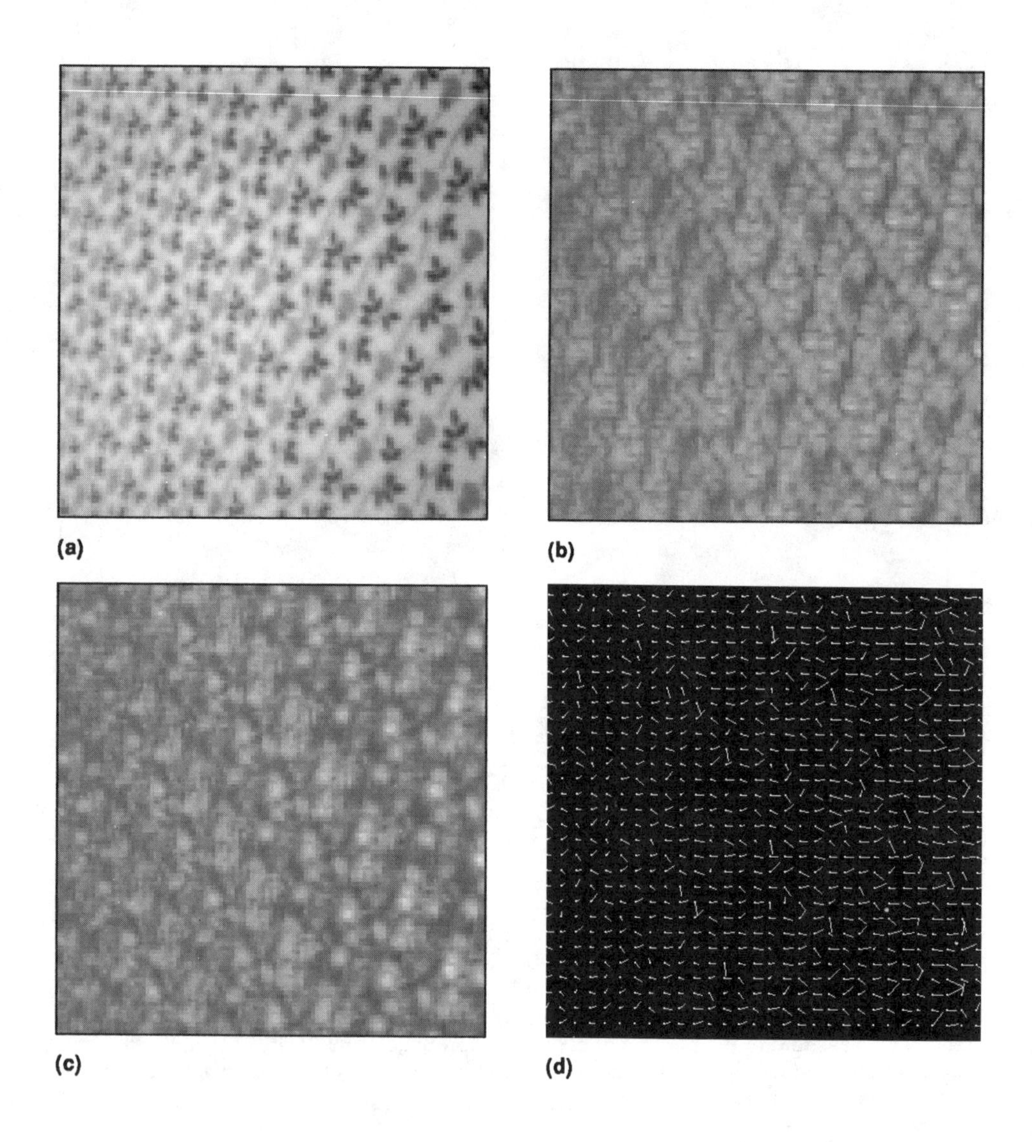

Figure 6.10. The poster experiment: (a) central frame of the image sequence, (b, c) confidence measures associated with conservation information, that is, the reciprocals of the eigenvalues of the covariance matrix S_{cc}, (d) initial estimate of velocity U_{cc}.

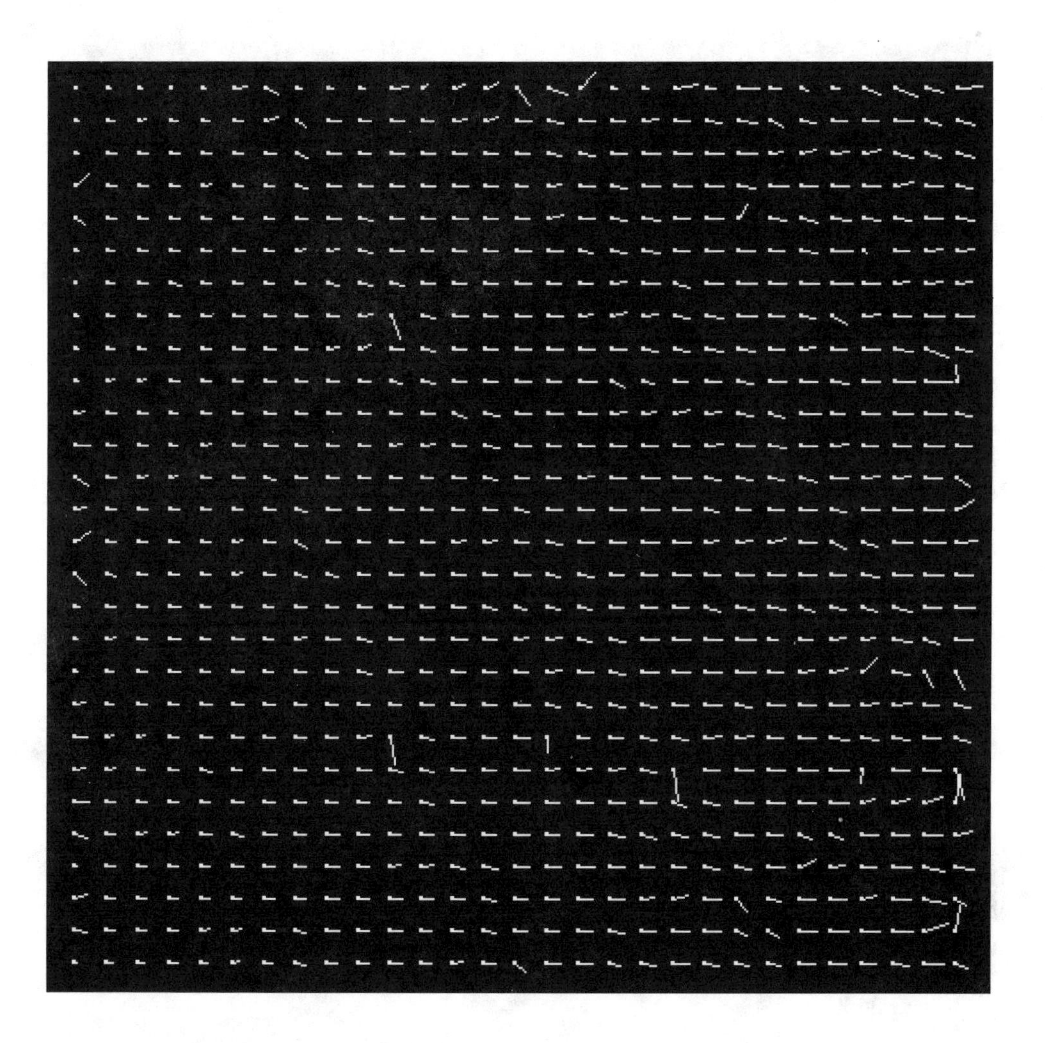

Figure 6.11. The poster experiment: flow field after velocity propagation (10 iterations).

2	2	2	3	8	5	5	4	11	5	8	9	11	8	5	12	9	12	7	6	10	15	12
3	5	5	5	6	2	8	7	8	8	8	9	10	9	10	10	11	11	10	10	10	11	10
6	6	5	5	5	6	8	8	8	8	8	7	9	9	11	11	11	12	10	11	11	13	11
5	6	6	5	6	6	8	8	8	9	9	9	9	10	10	10	11	10	10	12	11	11	10
2	5	6	6	6	6	7	8	8	8	8	9	9	8	10	10	11	11	11	13	14	10	11
2	5	6	6	6	6	6	8	7	9	8	10	8	9	10	10	12	11	11	12	5	11	11
3	4	5	6	6	7	6	8	8	9	8	9	9	10	9	10	11	11	10	9	10	13	12
4	4	5	6	6	6	6	7	7	8	8	9	9	10	7	8	10	10	11	11	11	15	11
6	5	5	6	6	7	8	7	7	8	9	10	9	11	8	9	10	9	12	12	10	12	12
7	5	5	6	7	7	7	7	7	8	9	9	8	10	10	11	10	10	10	10	10	11	11
6	5	5	5	6	7	8	7	8	7	8	9	10	9	10	10	8	9	11	10	11	12	11
8	6	6	5	6	6	8	7	9	7	8	9	9	9	10	11	9	10	11	9	10	14	14
6	6	6	5	6	6	7	7	8	8	8	9	9	8	9	11	10	11	11	10	11	12	11
3	5	7	6	6	6	7	8	8	9	8	7	9	9	10	10	10	11	8	9	10	11	11
3	5	6	6	6	6	7	8	8	8	9	9	9	9	11	11	11	12	9	11	10	10	3
6	5	6	7	6	7	6	7	7	8	8	9	8	9	9	10	12	11	10	12	11	11	9
7	4	6	7	6	7	6	8	6	9	8	10	9	8	10	9	11	10	10	12	10	11	11
7	5	5	6	6	7	7	7	8	8	7	9	9	10	10	2	11	11	12	13	14	11	13
7	5	5	6	7	7	8	7	7	8	8	9	9	10	9	9	10	10	9	8	10	12	11
3	5	5	6	7	7	7	7	8	8	8	9	9	11	8	8	10	10	11	9	10	13	12
7	6	6	5	6	7	8	7	8	8	8	8	8	10	9	10	10	23	11	11	11	16	4
4	5	5	5	6	7	8	7	9	6	8	8	9	9	9	10	10	8	11	11	11	14	10
	1	2	1	1	2	1	2	10	2	5	6	10	4	1	10	4	3	10	12	7	5	13

Figure 6.12. The poster experiment: magnitude of the horizontal component of the flow field after velocity propagation (multiplied by a scale factor of 10 and rounded to the nearest integer).

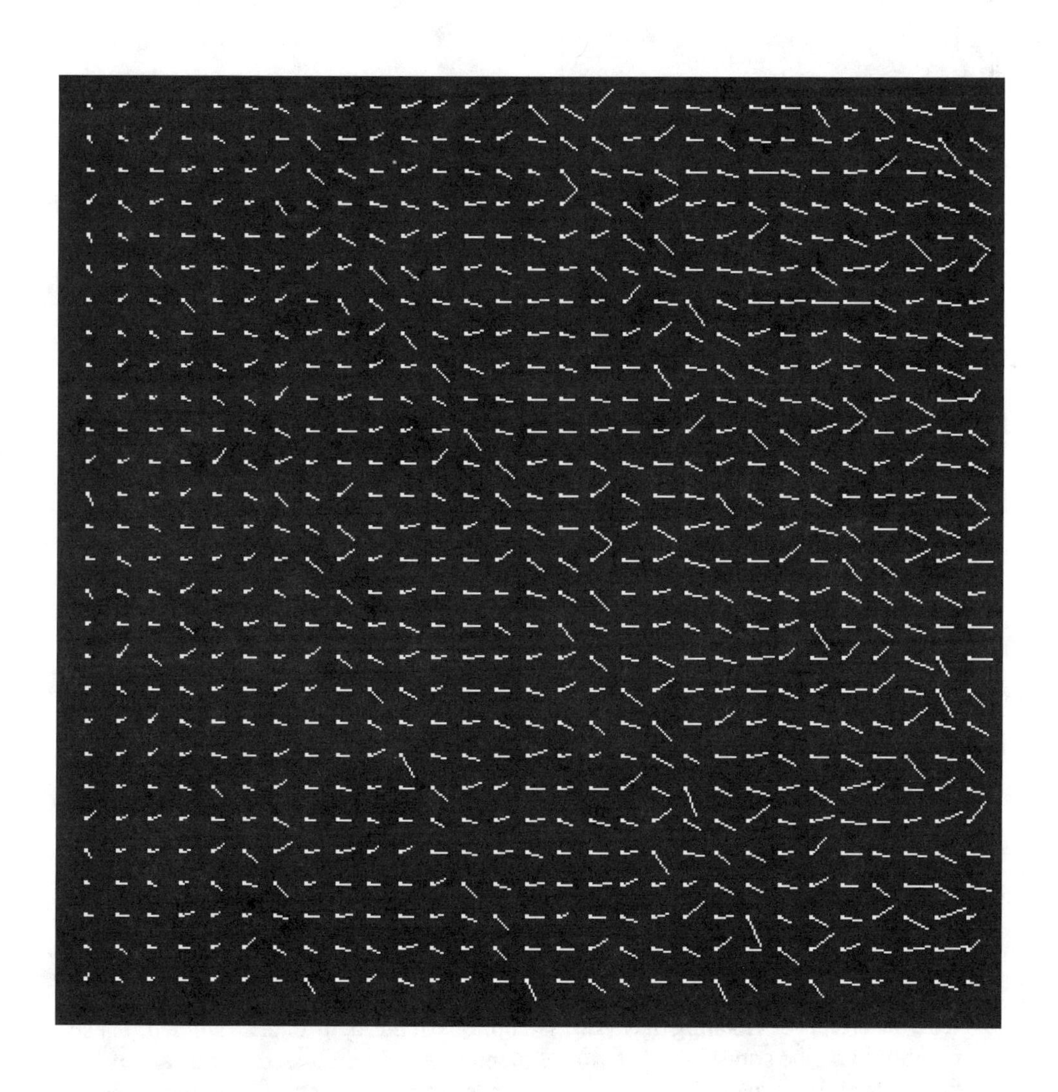

Figure 6.13. The poster experiment: flow field after 10 iterations of conventional smoothing (the algorithm converged in six iterations).

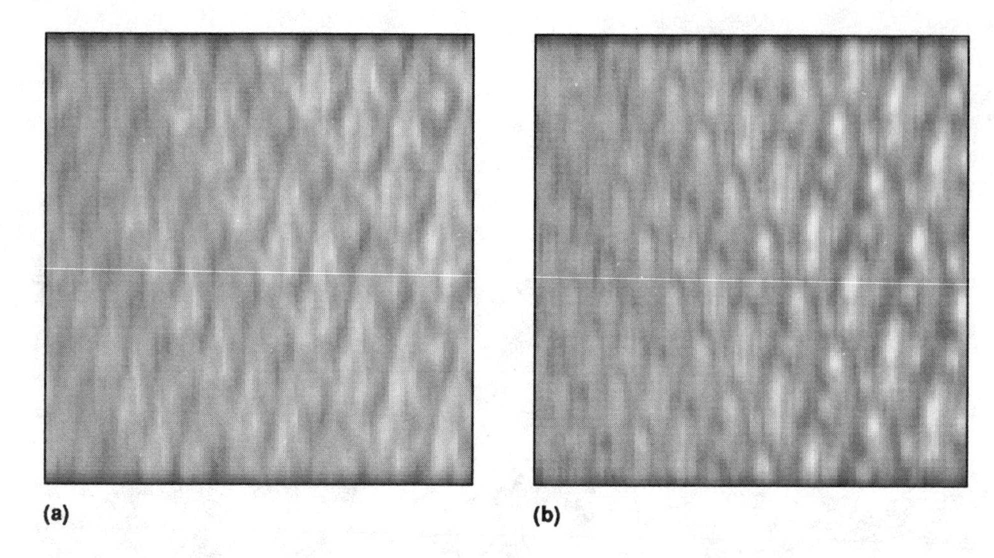

(a) (b)

Figure 6.14. The poster experiment: confidence measures associated with the flow field after velocity propagation.

One of the confidence measures is high both at edges and corners of the intensity image, whereas the other one is high only at corners. Figure 6.10d shows the initial estimate of the flow field (i.e., the velocity U_{cc}). Figure 6.11 shows the flow field after iterative velocity propagation (10 iterations). The flow field is qualitatively correct almost everywhere in the image, except at a few randomly placed points. The velocity estimate at these few points is incorrect because of a very high confidence associated with a wrong initial estimate (U_{cc}). As discussed earlier, such a situation can arise in some textured regions. Figure 6.12 shows the actual magnitude of the horizontal component of the flow field (scaled up by a factor of 10 and rounded to the nearest integer). The vertical component is zero and is not shown here. For comparison, Figure 6.13 shows the flow field after 10 iterations of conventional smoothing [34, 80] (with the smoothing factor α set to 0.5). Figures 6.14a and 6.14b show the two confidence measures after propagation.

Next, the effects of violating the basic assumptions of the algorithm are studied. These assumptions and the conditions that violate them have been discussed in the case of the Netherlands aerial imagery experiment. In the current experiment also, the first and the second assumptions are violated by setting the search window size to 3×3. The third assumption is violated by setting the search window size to 9×9 and using only two images to recover conservation information. Table 6.9 depicts the results and is self-explanatory. A comparison of the corresponding entries of Table 6.9 and Table 6.7 confirms the effects of violating the assumptions.

Table 6.9. Effects of violating the basic assumptions of the algorithm on the mean and standard deviation of the error in optic-flow components for the poster experiment. Postpropagation velocity is used in the computations.

Window Sizes	Horizontal Component		Vertical Component	
	Mean	Standard Deviation	Mean	Standard Deviation
3 × 3 Search	0.07	0.12	0.06	0.18
9 × 9 Search	0.12	0.08	0.10	0.11

The fused-image experiment. The first image for this experiment is generated by pasting together, side by side, a subimage of the first image in the aerial-imagery sequence and a subimage of the first image in the poster sequence. Figure 6.15a shows the pasted image. The second and third images are generated in a similar fashion, by using the second and third images of the two sequences, respectively. The ground-truth flow field can easily be generated by pasting together the corresponding subimages of the two flow fields shown in Figures 6.1b and 6.8b. Algorithm 1 is used to estimate optic flow. The sizes of the search window, correlation window, and velocity propagation neighborhood are set to 5 × 5, 3 × 3, and 3 × 3, respectively. Figures 6.15b and 6.15c show the initial confidence measures. Figure 6.15d shows the initial flow field U_{cc}. Especially notable are the velocity estimates in the vicinity of the semicircular boundary dividing the two original images. The aperture problem exists, although only to a small extent, in regions A and B. This is also confirmed by the two confidence measures—only one of them is high in these regions.

Figure 6.16 shows the flow field after velocity propagation (16 iterations). For comparison, Figure 6.17 shows the flow field obtained after 16 iterations of conventional smoothing [34, 80]. The motion boundary is much more crisp in Figure 6.16, despite the strong texture on either side of the boundary. To quantify the estimates of Figure 6.16, the percentages of pixels that have a magnitude of the vector error in velocity within 5 percent (of the true value), within 10 percent, and within 25 percent are determined in two regions of interest in the flow field. The first region is composed of a band of pixels whose shortest distance from the boundary is less than or equal to 10 pixels. The second region comprises the whole image. Table 6.10 shows the results. The performance at the boundary is very similar to the overall performance over the image. In other words, there is no degradation in performance specifically at the motion boundary.

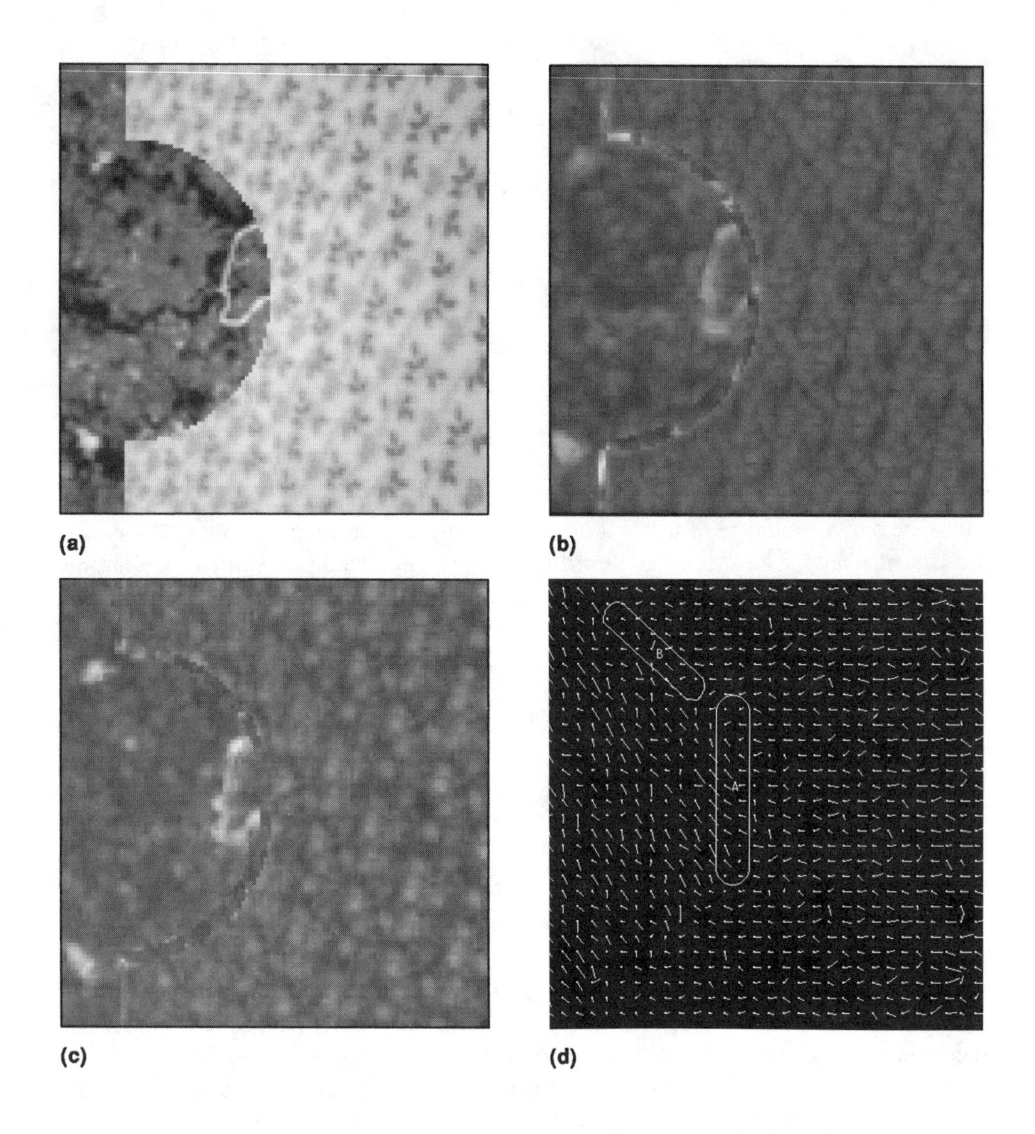

Figure 6.15. The fused-image experiment: (a) central frame of the image sequence, (b, c) confidence measures associated with conservation information, that is, the reciprocals of the eigenvalues of the covariance matrix S_{cc}, (d) initial estimate of velocity U_{cc}.

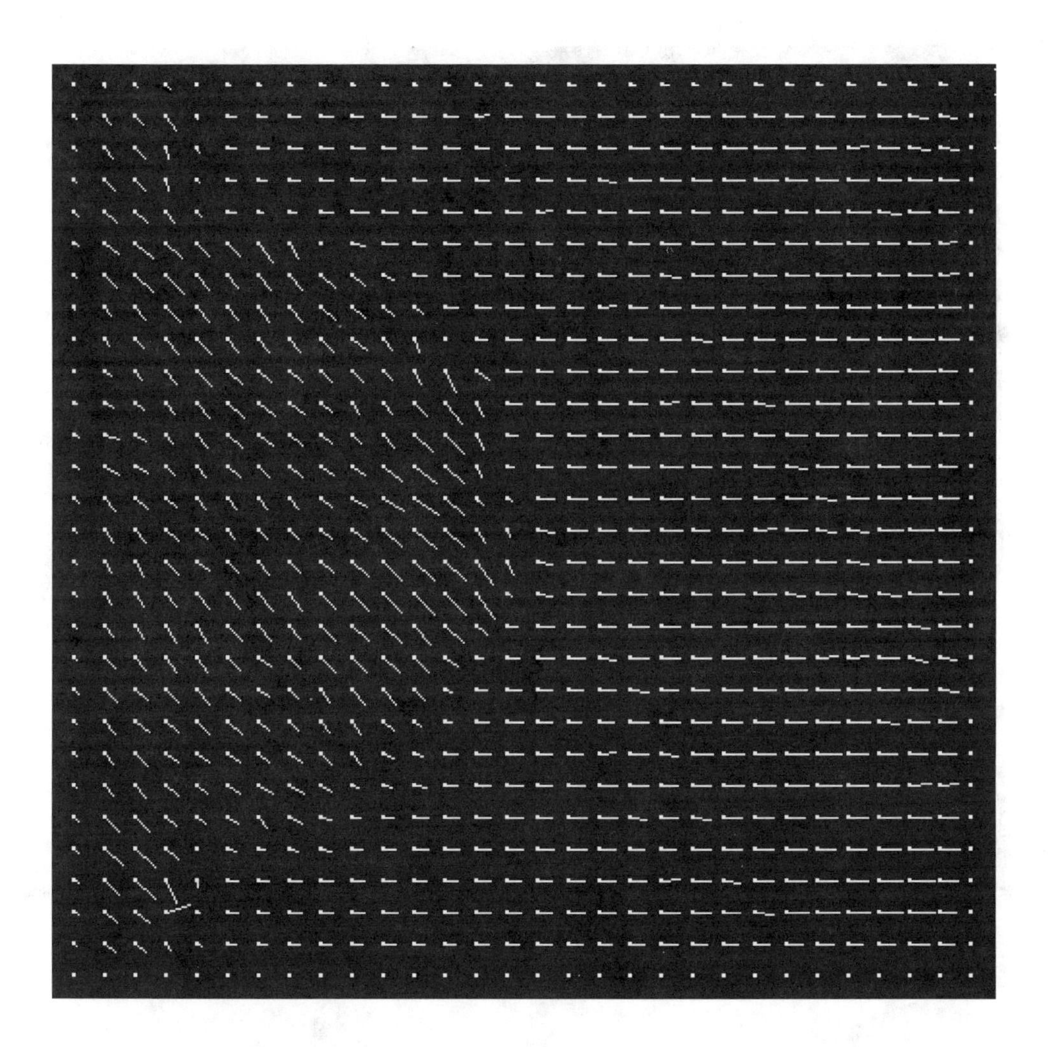

Figure 6.16. The fused-image experiment: flow field after velocity propagation (16 iterations).

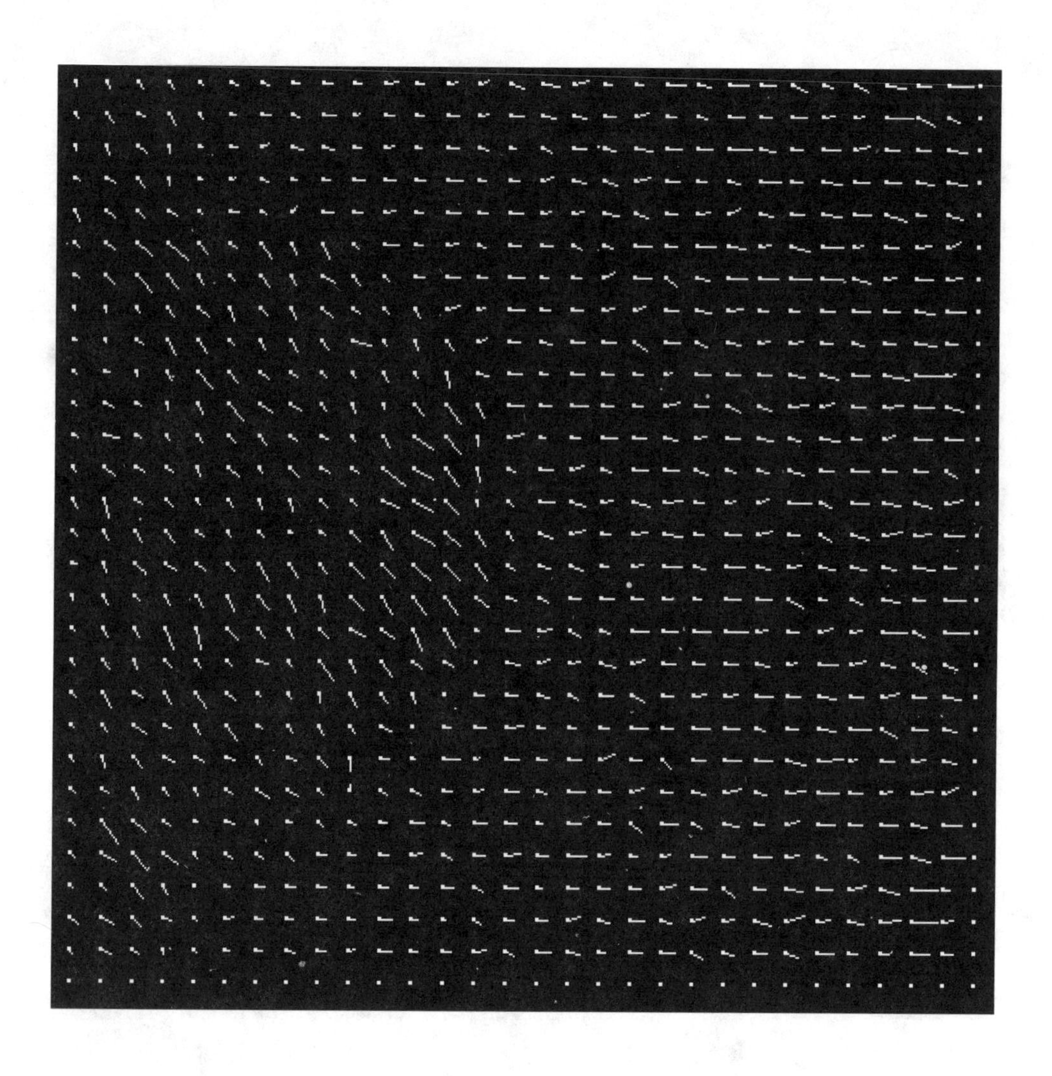

Figure 6.17. The fused-image experiment: flow field after 16 iterations of conventional smoothing.

Table 6.10. Error statistics for the fused-image experiment. The two rows correspond to the two regions of interest. For each row, the first and second columns indicate the percentages of total pixels for which the magnitude of the vector error in velocity is less than 5 percent before and after velocity propagation, respectively. The third and fourth columns give the corresponding percentages of pixels with error less than 10 percent. Finally, the fifth and sixth columns give the corresponding percentages of pixels with error less than 25 percent. The rows and columns closest to the image border are not used in the computation of these statistics.

Region of Interest	Percentage of pixels with vector error less than 5 percent		Percentage of pixels with vector error less than 10 percent		Percentage of pixels with vector error less than 25 percent	
	Without Propagation	With Propagation	Without Propagation	With Propagation	Without Propagation	With Propagation
Band of pixels around the motion boundary	61.0	65.7	72.6	81.5	75.5	92.3
Whole image	55.7	62.3	69.8	80.0	73.2	89.8

Qualitative experiments

The objective of this category of experiments is to judge the qualitative correctness of flow fields recovered by the algorithms, especially in terms of preservation of motion boundaries. A total of four experiments are conducted, with imagery corresponding to different scenes consisting of one or more moving objects against a stationary or moving background. In the first three experiments, the motion is known to be small. Hence, Algorithm 1 is used to estimate optic flow. In the last experiment, the motion can range from very small to very large. Hence, Algorithm 2 is used. In most of the experiments, the images are acquired using a 256×242 camera, and the maximum image motion is about three pixels per frame. For optic-flow computations, the images are low-pass filtered and subsampled to get a resolution of 128×121. At this level of resolution, the maximum optic flow is expected to be about one and a half pixels per frame. In all the experiments, the sizes of the correlation window, search window, and propagation neighborhood are 3×3, 5×5, and 3×3, respectively. Three frames are used for recovery of conservation information (except in the last experiment, where only two frames are used) and velocity propagation is carried out for 10 iterations. In the various flow-field images that follow, the velocity vector for only every fourth pixel (in both the horizontal and vertical directions) is shown for clarity. Further, the magnitude of velocity is multiplied by a scale factor of four to make the velocity vector clearly visible.

The toy truck experiment. The first experiment uses a toy truck on a flat (and mostly dark) table. Three images are shot as the truck rolls forward. The motion is largely translational, except in the vicinity of the wheels, where it has a small rotational component. Furthermore, the motion boundaries are expected to show up primarily as step discontinuities in the flow field. Figures 6.18 through 6.22 show various flow fields and confidence measures. Figure 6.18a shows the central frame of the original sequence. Figures 6.18b and 6.18c show the two confidence measures associated with conservation information at each point in the visual field. One of the confidence measures is high both at edges and corners of the intensity image, whereas the other is high only at corners. Figure 6.18d shows the initial estimate of the flow field (i.e., the velocity U_{cc}). Figure 6.19 shows the flow field after iterative velocity propagation (10 iterations), superimposed on the wireframe of the truck. The wireframe is obtained using a Canny-like edge detector. Figure 6.20 shows the flow field after 10 iterations of conventional smoothing [34, 80] (with the smoothing factor α set to 0.5), also superimposed on the wireframe.

A comparison of Figures 6.19 and 6.20 clearly shows that the propagation procedure developed in the new framework does an excellent job of preserving motion boundaries. There is very little "bleeding" of velocity from the truck into the background in Figure 6.19. On the other hand, there is considerable blurring of motion boundaries in Figure 6.20. For illustration, the actual magnitude of the vertical component of the flow field of Figure 6.19 is shown in Figure 6.21 (the values less than 0.1 are not shown to enhance clarity). The horizontal component is similar and is not shown here. Figures 6.22a and 6.22b show the two confidence measures after propagation. As expected, the confidence has propagated outward from the prepropagation high-confidence regions.

The tori experiment. The second experiment deals with two moving objects. Here, the camera is looking at two tori, one brighter than the other. Three images are shot as the brighter torus moves to the left (in the image) and the darker torus moves upward and to the left. Figures 6.23 through 6.27 show various flow fields and confidence measures. The figures are self-explanatory. Particularly notable is the flow field shown in Figure 6.24. A well-defined "hole" in the flow field corresponding to each torus clearly shows the boundary-preserving feature of the algorithm. On the other hand, in Figure 6.25 (where the flow field is obtained by conventional smoothing) the pixels in the hole (that actually belong to the background) have noticeable velocities. This is because of the blurring introduced by conventional smoothing.

The block-on-the-sweater experiment. The third experiment uses images (taken by a stationary camera) of a wooden block placed on a textured sweater. Only the block moves (to the left and upward). In other words, this is a setting where an object is moving against a textured rather than a flat background. Furthermore, the camera is placed very close to the block to introduce some perspective effects. This, however, has led to blurring of intensity edges due to defocusing.

Figures 6.28 through 6.32 show various flow fields and confidence measures. Figure 6.28a shows the central frame of the original sequence. Figures 6.28b and 6.28c show the two confidence measures associated with conservation information at each point in the visual field. As expected, one of the confidence measures is high both at edges and corners of the intensity image, whereas the other is high only at corners. Whether a point lies on the moving object or the stationary background does not affect the nature of conservation confidence measures. Thus, all edges and corners—irrespective of whether they correspond to surface discontinuities of the block or texture markings on the sweater—get a high confidence in Figures 6.28b and 6.28c, respectively. Figure 6.28d shows the initial estimate of the flow field (i.e., the velocity U_{cc}). Figure 6.29 shows the flow field after iterative velocity propagation (10 iterations), superimposed on the wireframe of the block. The perspective effect is quite marked. It shows up as a change in the direction of velocity vectors in the regions where two surfaces meet (especially at the intersection of the surface with two holes and the shiny surface).

For comparison, Figure 6.30 shows the flow field after 10 iterations of conventional smoothing [34, 80] (with the smoothing factor α set to 0.5), also superimposed on the wireframe. The difference between Figures 6.29 and 6.30, although noticeable, is negligible. This is because the edges of the block in the intensity image are too blurred to lead to crisply defined motion boundaries in the flow field. Furthermore, the background is textured and not flat. Also, neighborhood velocity distribution for the pixels lying on motion boundaries is not strictly "two-clustered" during the iterative update procedure. For illustration, the actual magnitude of the vertical component of the flow field of Figure 6.29 is shown in Figure 6.31. Figures 6.32a and 6.32b show the two confidence measures after propagation.

The dinosaur experiment. The last experiment uses a sequence of two 128×128 images of a toy dinosaur and a tea box on a tiled table surface against a textured background. (This sequence was shot at the University of Massachusetts and only two images are available.) Figure 6.33 shows the image pair. The 3D motion between the two frames consists of a translation of the camera to the right along with a leftward rotation about the vertical axis (to bring the scene back in view), as well as an independent leftward movement of the dinosaur. The leftward rotation of the camera is expected to induce a rightward flow in the image. However, everywhere except in the background and on the table, it is compensated for by the rightward motion of the camera and the motion of the dinosaur. As a result, the optic flow in the background is rightward and that on the table is almost zero. On the other hand, the optic flow on the tea box and the dinosaur is leftward. The motion is large enough to require Algorithm 2. The ground-truth flow field is not known, however.

Burt's Laplacian pyramid is constructed from each of the images. Four levels of the pyramid are sufficient. Figures 6.34a, 6.34b, 6.34c, and 6.34d show the first frame at level 3, level 2, level 1, and level 0 of the pyramid, respectively.

Figures 6.35 and 6.36 show the flow fields before and after propagation, respectively, at the coarsest level (16×16). The postpropagation flow field has the correct general direction everywhere, but the table seems to be influenced by the dinosaur. This is because the table has practically no intensity variation at this level of resolution, resulting in a very low prepropagation confidence. During propagation, the dinosaur flow dominates because it has a higher prepropagation confidence. This causes the table to acquire the same general flow as the dinosaur.

Figures 6.37 through 6.42 depict the flow fields before and after propagation at level 2, level 1, and level 0, respectively. For clarity, only a 32×32 sample of the flow field is shown at level 1 and level 0. Figure 6.43 shows the same flow field as Figure 6.42, except that certain interesting regions are labeled for further explanation. Also, Figure 6.44 shows the flow field at level 0, obtained after 10 iterations of conventional smoothing [34, 80]. The motion in this experiment is rather complex. The flow field recovered from this sequence reveals several positive and negative aspects of the algorithm.

Positives. Region C in Figure 6.43 corresponds to a step discontinuity in the flow field. It is different from the discontinuities in the previous experiments because it is caused by motion of the object and the background in opposite directions. A comparison of Figures 6.43 and 6.44 clearly reveals that the discontinuity has been preserved very well by the new algorithm. At the shown resolution, there is only one line of pixels in C that has an almost zero flow. On the two sides of this line, the pixels have almost correct velocities (of the background and the dinosaur, respectively). Region D has a similar discontinuity. Again, the flow from the dinosaur has bled much less into that of the table as compared with Figure 6.44. Although there is considerable blurring in region D at low resolutions, the discontinuity is very crisp at the highest resolution. This is because the intensity discontinuities are more prominent at higher resolutions. This leads to a high confidence in conservation-based estimates (U_{cc}) near discontinuities (note that the motion discontinuities coincide with intensity discontinuities in this case) that prevents excessive smoothing.

Negatives. Although the blurring of the flow field is less prominent at discontinuities (such as in region D) as compared with conventional smoothing, it still exists. This is unlike most of the discontinuities in the last six experiments where the blurring is very negligible. A possible reason is that the change in velocity across the discontinuity is very high in the current experiment. Also, in region G, although the velocity change across the discontinuity is not very large, the optic flow has still bled beyond the left edge of the tea box, up to the closest vertical edge of the tile. This is because the two vertical features (the box edge and the tile edge) are so close together that they have merged at level 4 and level 3 of the pyramid. The region between the two edges, therefore, is influenced by the motion of the box at these levels of resolution and is assigned incorrect velocity. This incorrect velocity is propagated to the finer resolutions.

The velocity estimates in regions A, B, and F are inconsistent with their immediate neighborhoods. Region A has a high-frequency texture (not visible at the dithered gray-scale resolution in the pictures shown here) that leads to an incorrect conservation estimate with a high confidence that cannot be corrected during propagation. Region B corresponds to a neighborhood that spans a discontinuity. Pixels from one side of the discontinuity (the dinosaur side) have a strong rotational component in their velocity that leads to a "scattered" neighborhood velocity distribution. This leads to a very low confidence in neighborhood information, making velocity propagation ineffective. Region F belongs to the table, where the rightward translation and the leftward rotation of the camera almost cancel out. The spatial resolution seems insufficient to measure this velocity correctly.

Conclusion

I have attempted to do a fairly exhaustive experimental evaluation of the framework in this chapter. The two categories of experiments span a gamut of situations in terms of the nature of the moving objects and the background. While the first category focuses on the quantitative correctness of the flow fields recovered by the new algorithm, the second category examines their qualitative behavior—at motion boundaries as well as in flat regions. Various tests have been done to evaluate the robustness of the framework to the addition of noise. On the whole, the framework performs very well. It does a very good job of preserving the step discontinuities in the flow field when there is no texture in the vicinity of discontinuities and the change of velocity across the discontinuity is not very large.

The framework has been tested primarily at a qualitative level in situations where the whole visual field is not moving (that is, in situations where one or more objects are moving against a stationary or moving background and the resulting flow field is not continuous). The only quantitative experiment involving a motion discontinuity described here uses "quasi-real" imagery, where the discontinuity is generated by pasting together two real images. This is because there is no straightforward way to find out the ground-truth optic flow when the discontinuity is caused by an object moving against a background. Recovery of even an "almost correct" flow field by the back-projection scheme used for the poster experiment would require an accurate depth map and correct image segmentation.

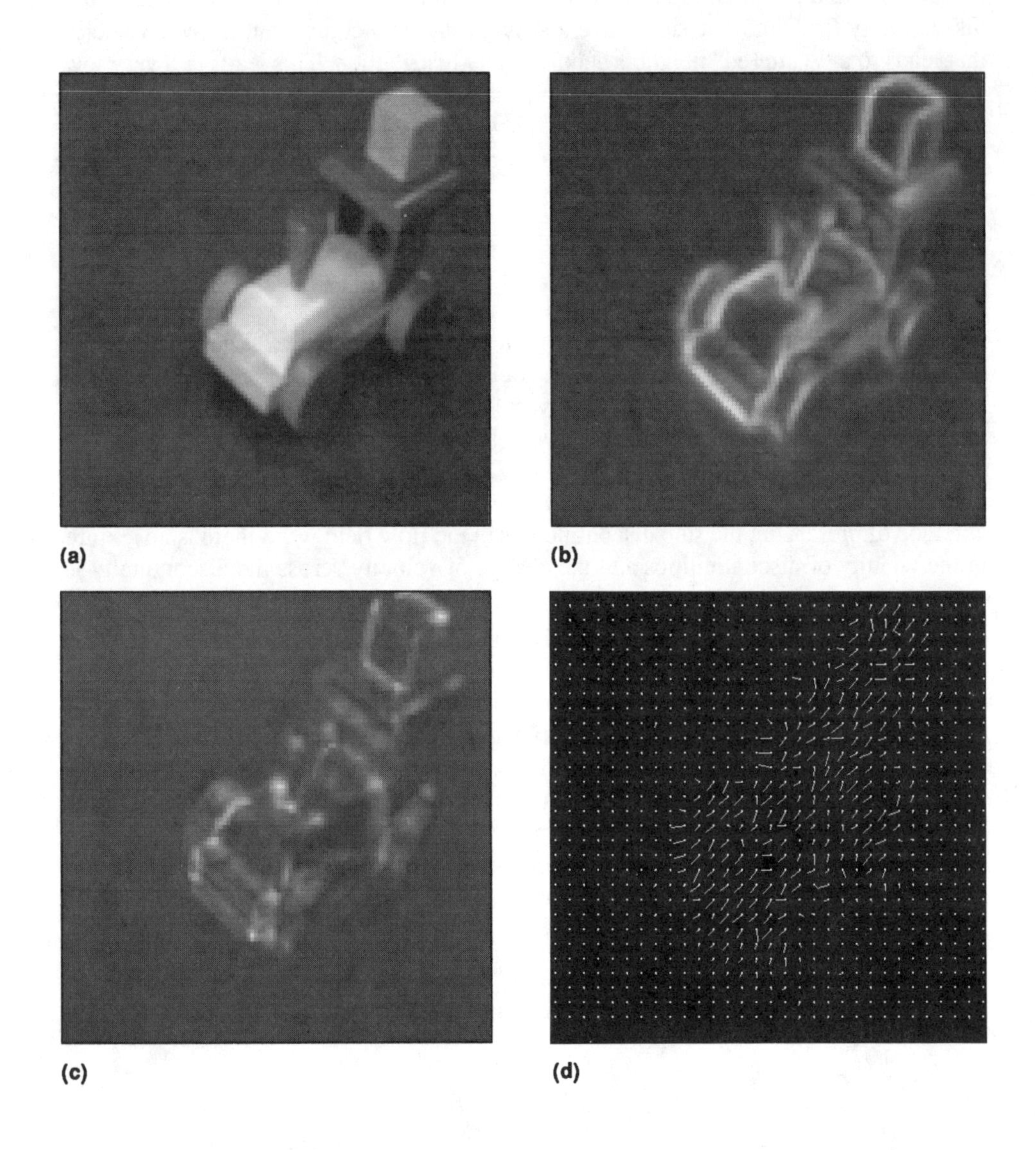

(a) (b) (c) (d)

Figure 6.18. The toy truck experiment: (a) central frame of the image sequence, (b, c) confidence measures associated with conservation information, that is, the reciprocals of the eigenvalues of the covariance matrix S_{cc}, (d) initial estimate of velocity U_{cc}.

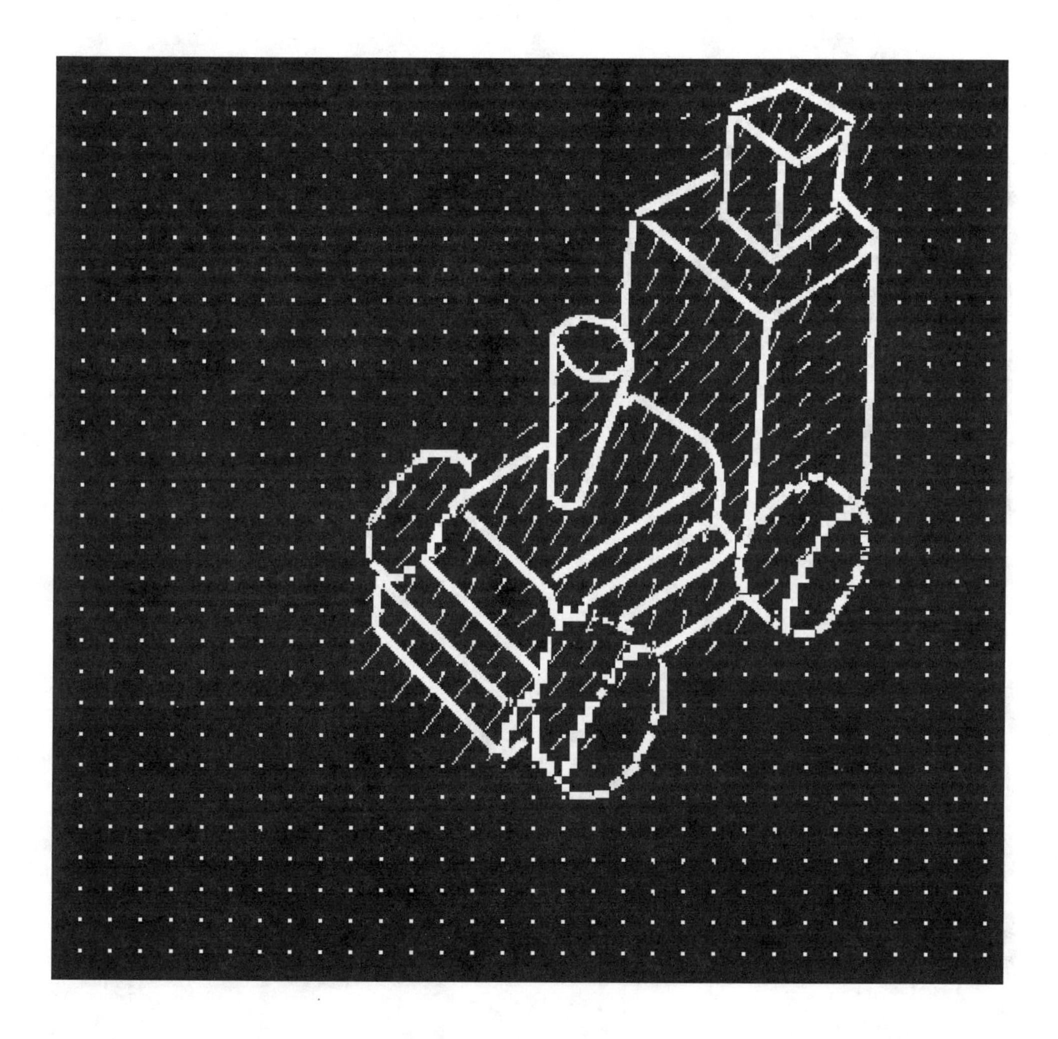

Figure 6.19. The toy truck experiment: flow field after velocity propagation, superimposed on the wireframe of the truck.

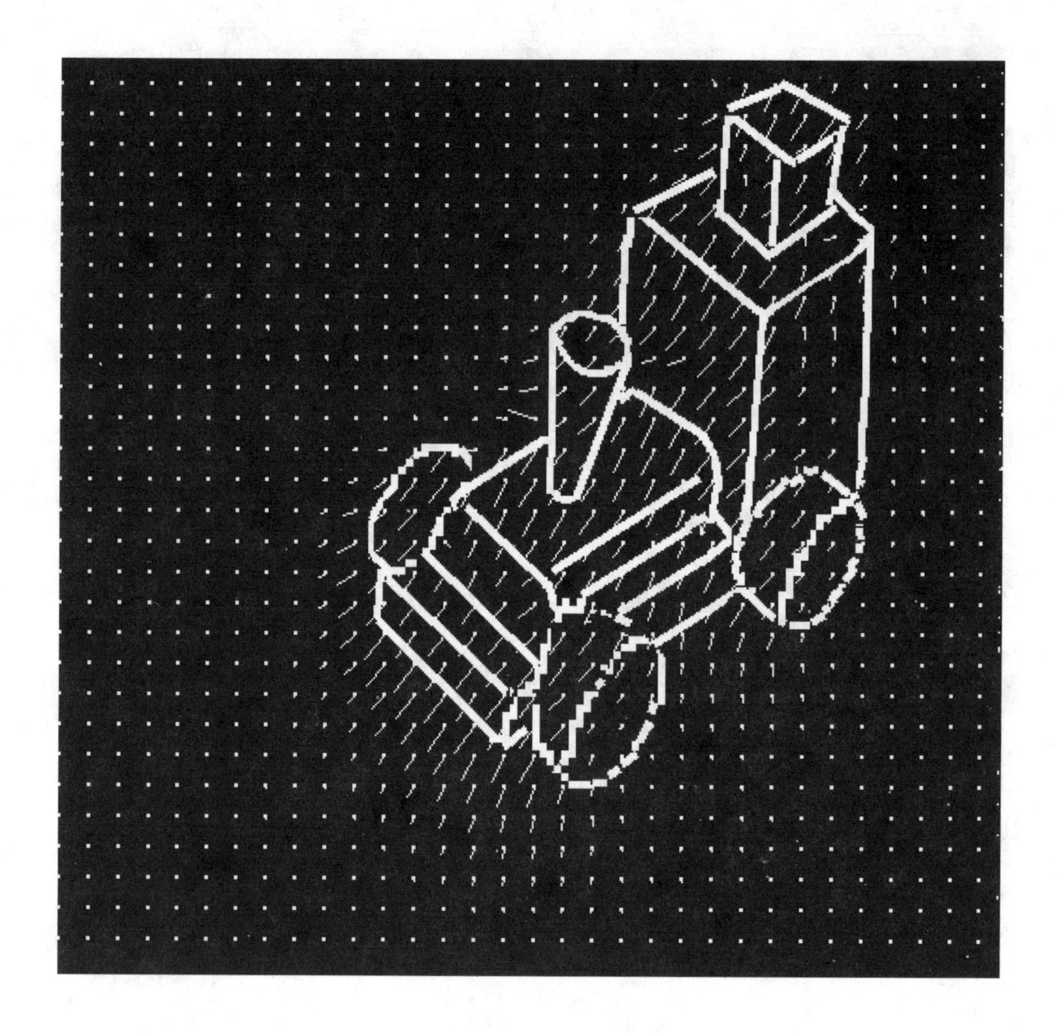

Figure 6.20. The toy truck experiment: flow field after 10 iterations of conventional smoothing, superimposed on the wireframe of the truck.

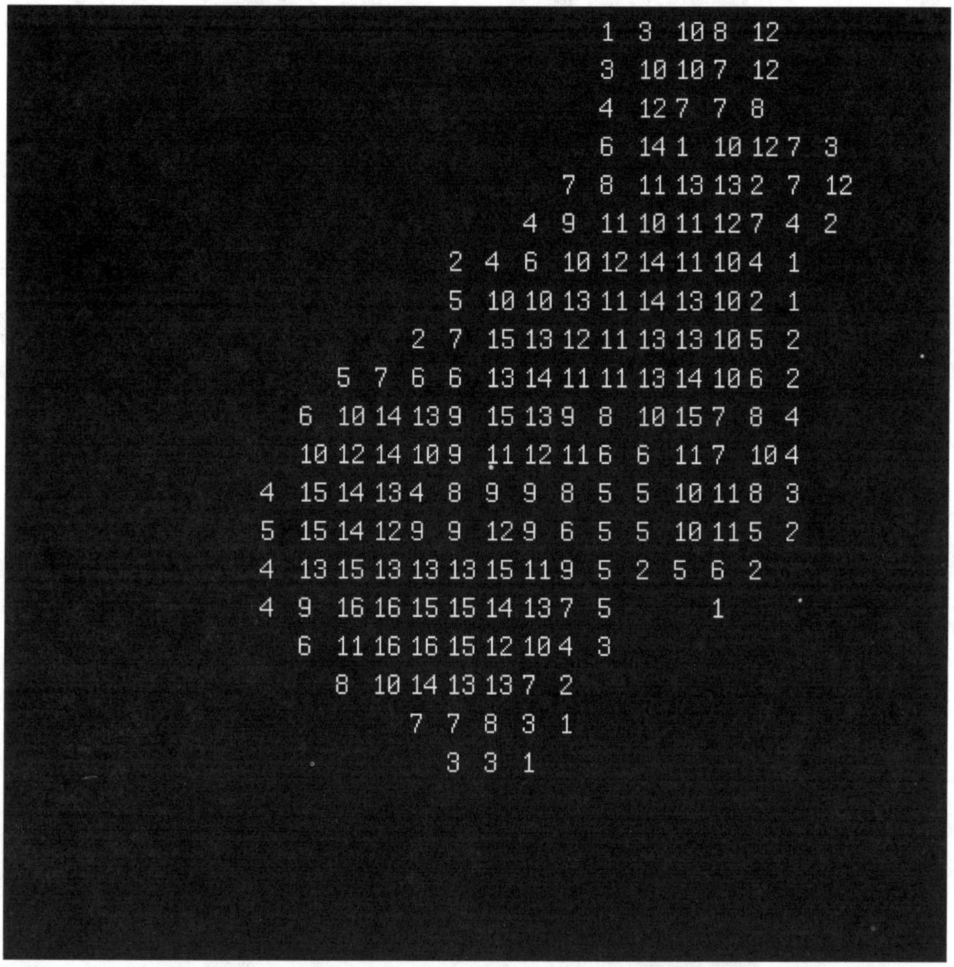

```
                                1  3  10 8  12
                                3  10 10 7  12
                                4  12 7  7  8
                                6  14 1  10 12 7  3
                             7  8  11 13 13 2  7  12
                          4  9  11 10 11 12 7  4  2
                    2  4  6  10 12 14 11 10 4  1
                    5  10 10 13 11 14 13 10 2  1
                 2  7  15 13 12 11 13 13 10 5  2
           5  7  6  6  13 14 11 11 13 14 10 6  2
        6  10 14 13 9  15 13 9  8  10 15 7  8  4
        10 12 14 10 9  11 12 11 6  6  11 7  10 4
     4  15 14 13 4  8  9  9  8  5  5  10 11 8  3
     5  15 14 12 9  9  12 9  6  5  5  10 11 5  2
     4  13 15 13 13 13 15 11 9  5  2  5  6  2
     4  9  16 16 15 15 14 13 7  5        1
        6  11 16 16 15 12 10 4  3
           8  10 14 13 13 7  2
                 7  7  8  3  1
                    3  3  1
```

Figure 6.21. The toy truck experiment: magnitude of the vertical component of the flow field after velocity propagation (multiplied by a scale factor of 10 and rounded to the nearest integer). The values less than 1 are not displayed to enhance clarity.

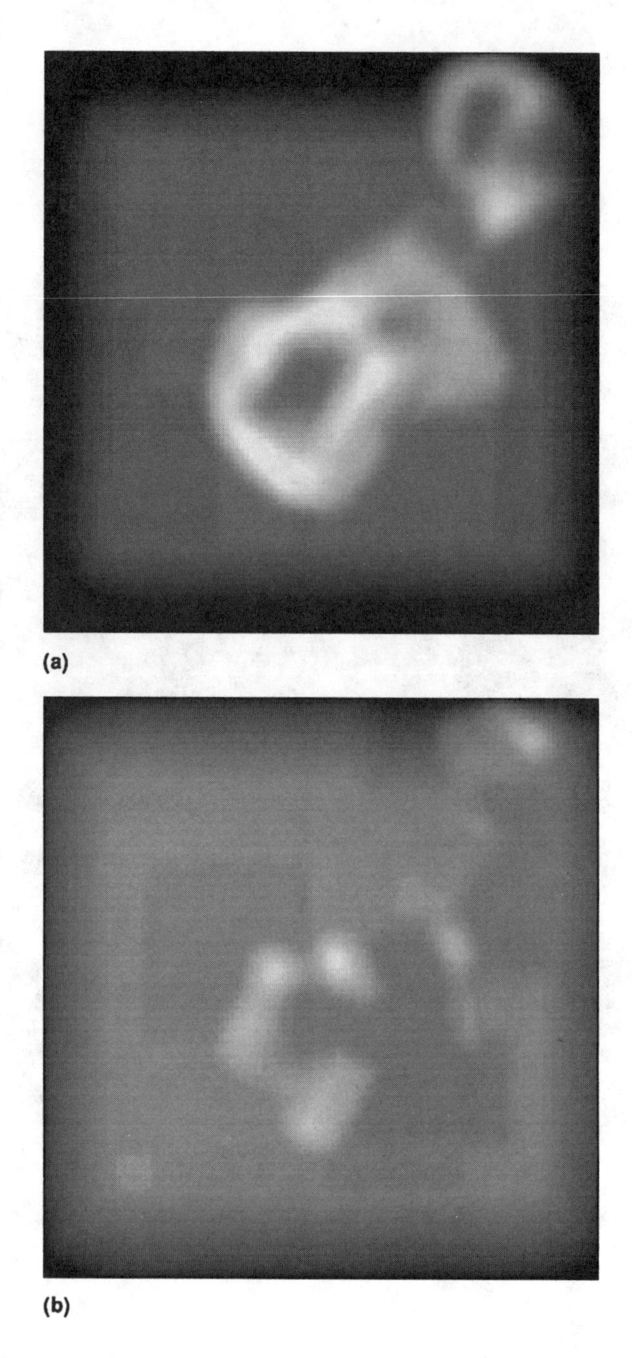

Figure 6.22. The toy truck experiment: confidence measures associated with the flow field after velocity propagation.

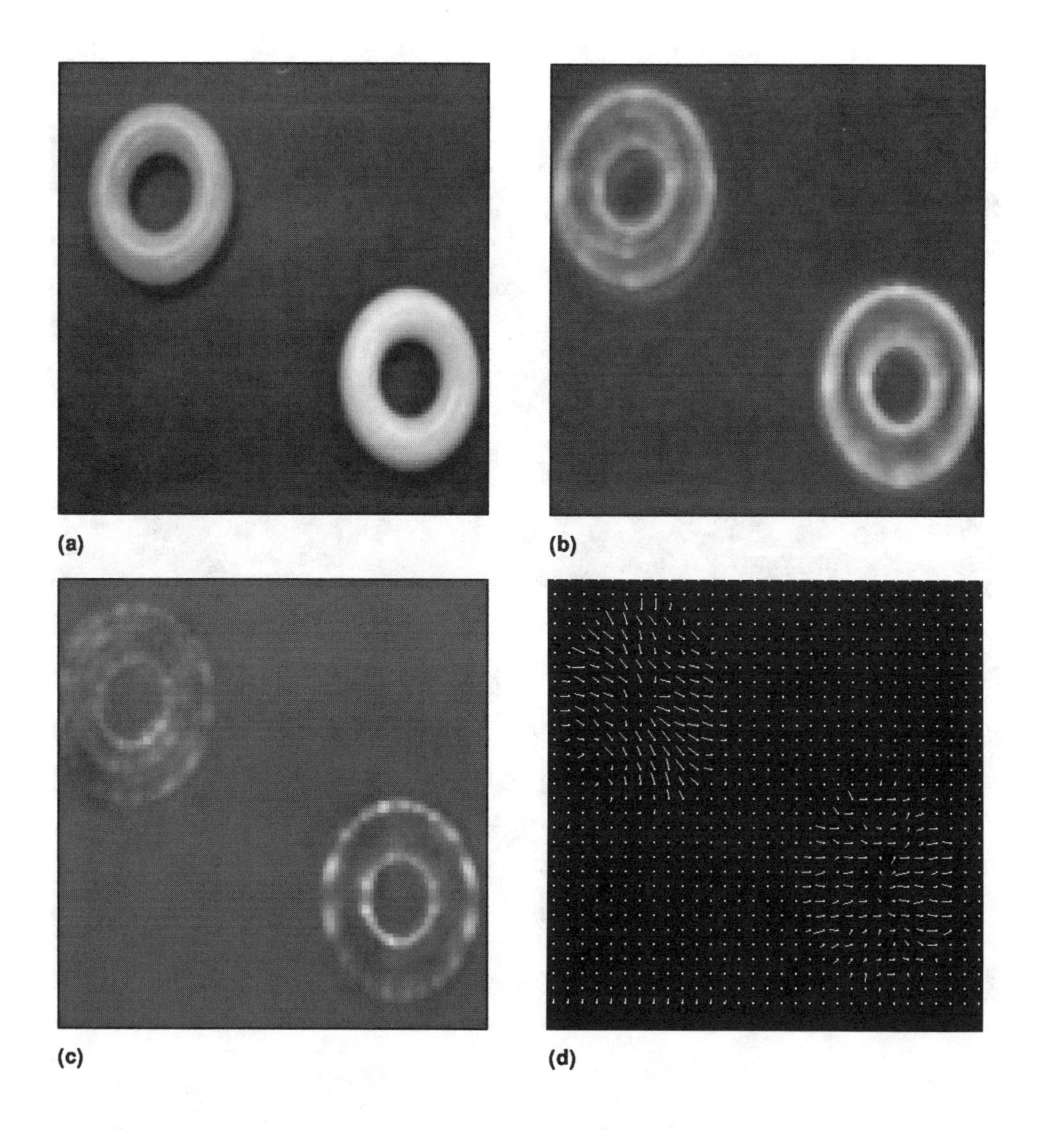

Figure 6.23. The tori experiment: (a) central frame of the image sequence, (b, c) confidence measures associated with conservation information, that is, the reciprocals of the eigenvalues of the covariance matrix S_{cc}, (d) initial estimate of velocity U_{cc}.

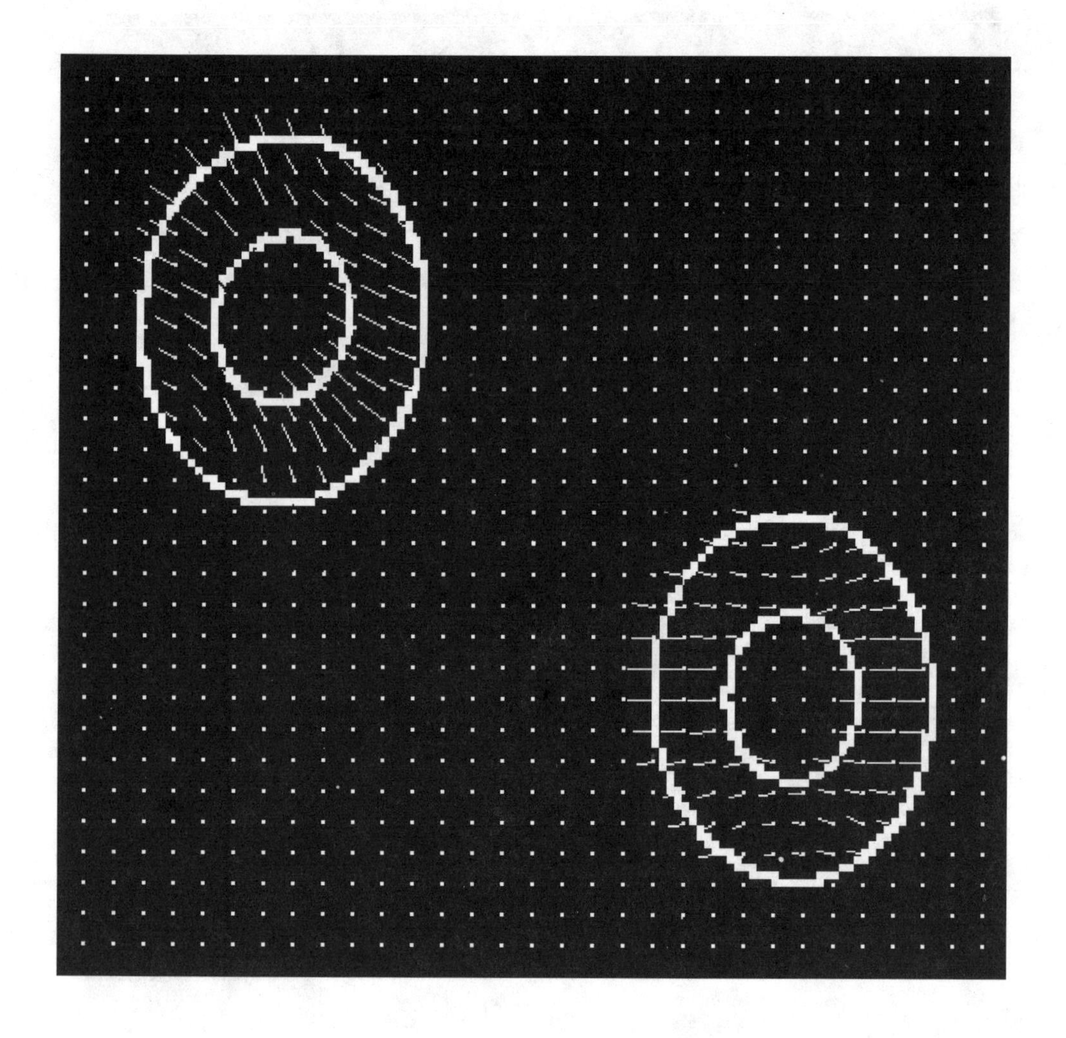

Figure 6.24. The tori experiment: flow field after velocity propagation, superimposed on the wireframes of the tori.

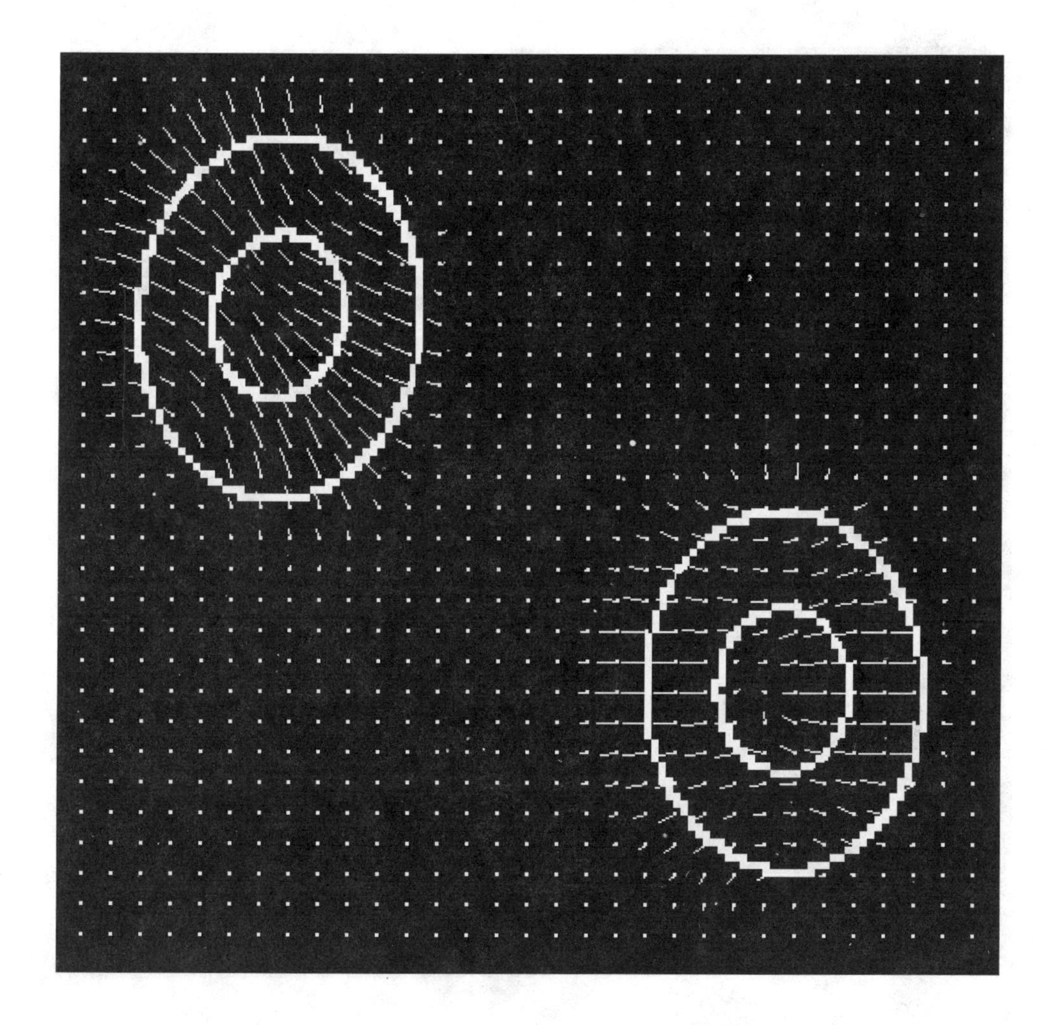

Figure 6.25. The tori experiment: flow field after 10 iterations of conventional smoothing, superimposed on the wireframes.

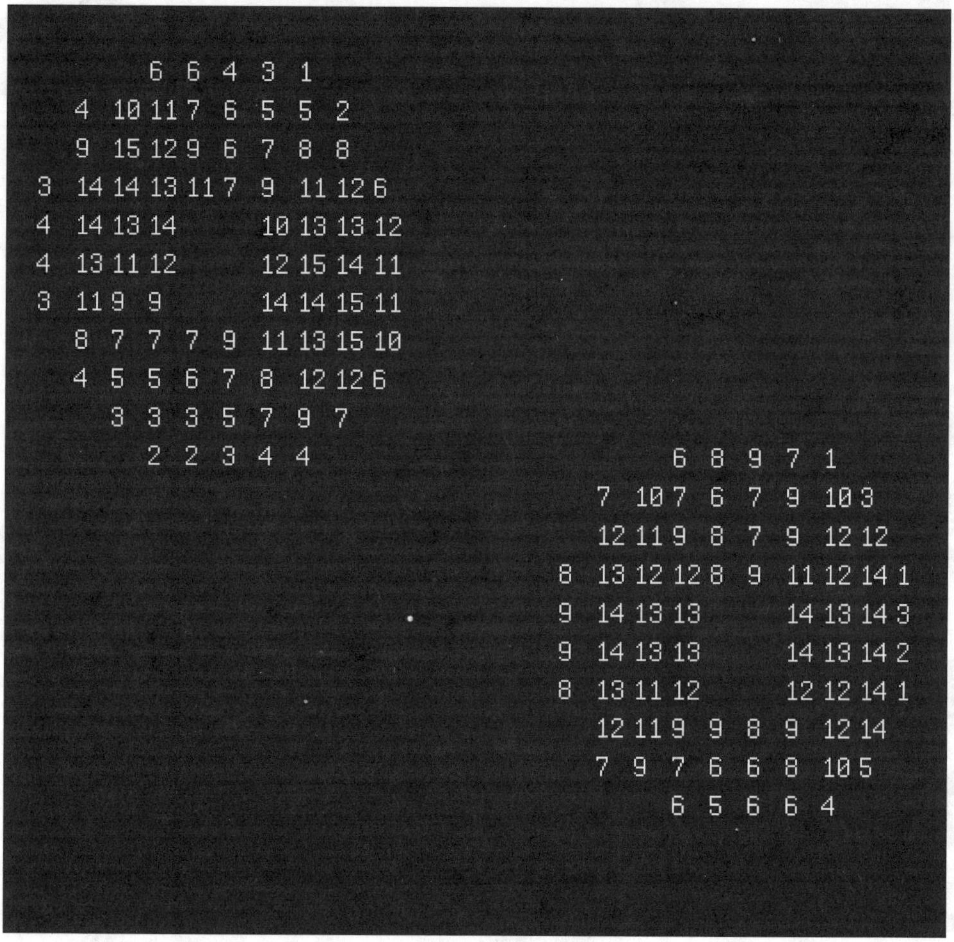

```
      6  6  4  3  1
   4  10 11 7  6  5  5  2
   9  15 12 9  6  7  8  8
3  14 14 13 11 7  9  11 12 6
4  14 13 14       10 13 13 12
4  13 11 12       12 15 14 11
3  11 9  9        14 14 15 11
   8  7  7  7  9  11 13 15 10
   4  5  5  6  7  8  12 12 6
      3  3  3  5  7  9  7
         2  2  3  4  4
                                    6  8  9  7  1
                              7  10 7  6  7  9  10 3
                              12 11 9  8  7  9  12 12
                           8  13 12 12 8  9  11 12 14 1
                           9  14 13 13       14 13 14 3
                           9  14 13 13       14 13 14 2
                           8  13 11 12       12 12 14 1
                              12 11 9  9  8  9  12 14
                              7  9  7  6  6  8  10 5
                                    6  5  6  6  4
```

Figure 6.26. The tori experiment: magnitude of the horizontal component of the flow field after velocity propagation (multiplied by a scale factor of 10 and rounded to the nearest integer).

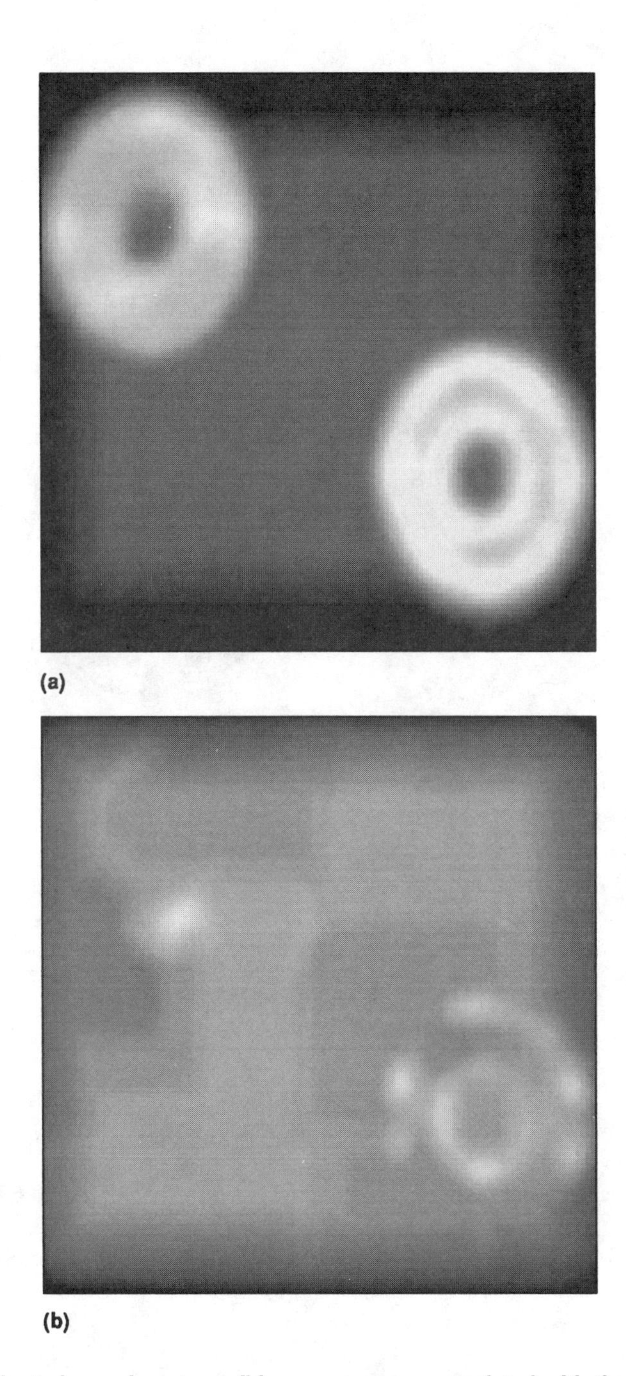

Figure 6.27. The tori experiment: confidence measures associated with the flow field after velocity propagation.

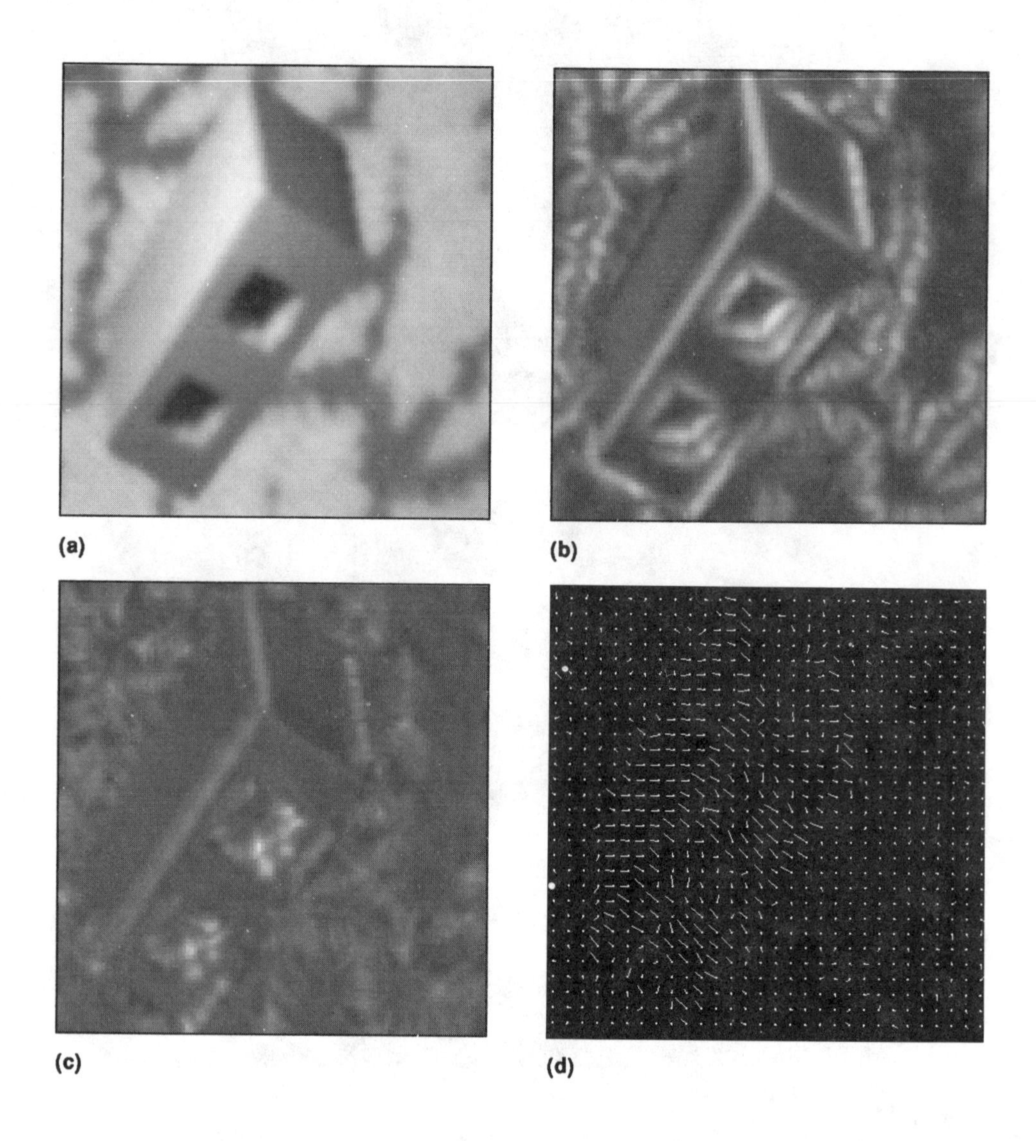

Figure 6.28. The block-on-the-sweater experiment: (a) central frame of the image sequence, (b, c) confidence measures associated with conservation information, that is, the reciprocals of the eigenvalues of the covariance matrix S_{cc}, (d) initial estimate of velocity U_{cc}.

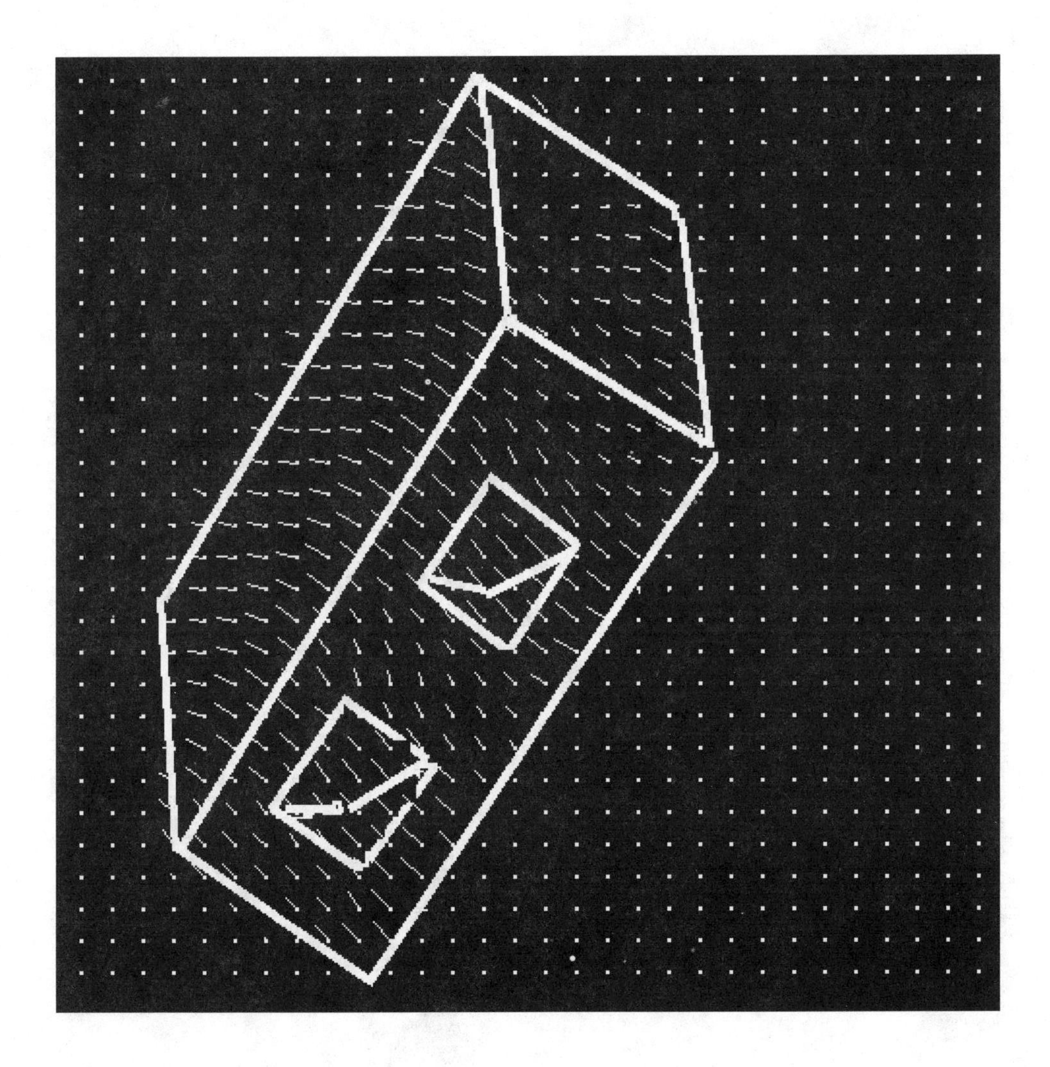

Figure 6.29. The block-on-the-sweater experiment: flow field after velocity propagation, superimposed on the wireframe of the block.

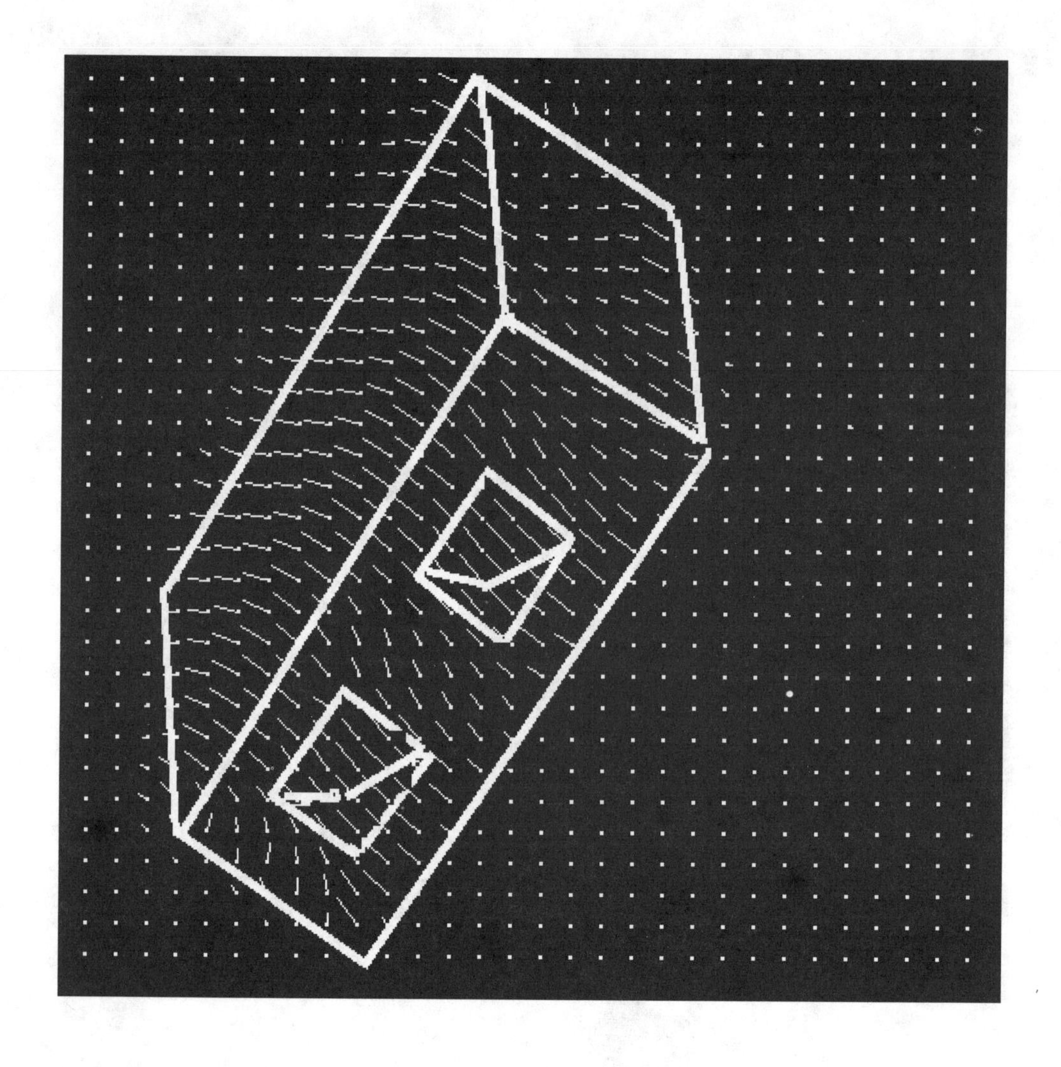

Figure 6.30. The block-on-the-sweater experiment: flow field after 10 iterations of conventional smoothing, superimposed on the wireframe.

```
                       1  3  8     2  1
                    1  1  5 11    16  4  2
                    3  1  4 10  2  3  3  1  1
                 2  1  2  3  8  2  4  5  1  5
                 1  1  2  2  6  6  1  5  2  9  1
              2  1  2  2  3  6  9  4  1  2  7  1
           1  2  2  2  2  5 10  9  7  4  3  6  4
           2  4  3  2  4  8 11  8  6  6  6  7 11
        1  4  3  2  3  6 12 10  8  6  5  5  8 10  1
        3  4  3  2  5 10 12 13 10  8  6  5  5  3  1
     1  4  4  2  3  8 13 14 12 11 12  9  6  5  2  1
     3  3  3  3  6 11 12 13  8 10 15 13 10  4  1
     5  2  2  4  9 11 12 13 14 17 15 13 13  2
     5  1  3  7 12 10  9 12 16 16 15 14  4  1
  1  6  2  5 11 11 10  8  9 11 15 13  7  1
  1  8  3  9 12 13 12 10  9  8  9  7  1
  1  7  5 11 12  9  5 11 11  9  7  2
  2  2 10 10 12 12 13 16 13 10  4  1
  2  8 11  9 12 16 16 16 12  6  1
  2  9 11  8  8 11 14 15 14
     2  5  9 11  9 11 15  1
        1  4  7 11 16 15
                  12
```

Figure 6.31. The block-on-the-sweater experiment: magnitude of the vertical component of the flow field after velocity propagation (multiplied by a scale factor of 10 and rounded to the nearest integer).

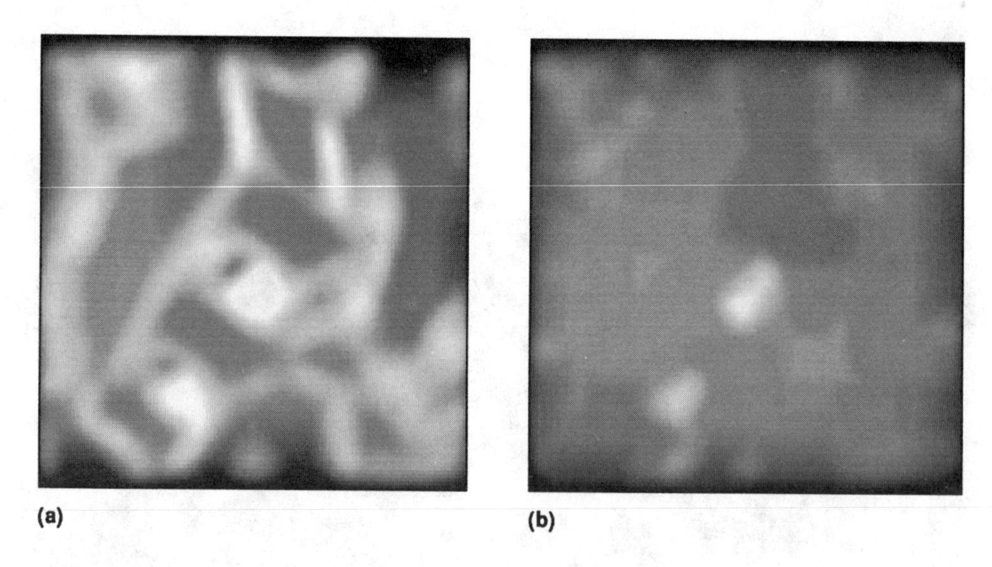

(a) (b)

Figure 6.32. The block-on-the-sweater experiment: confidence measures associated with the flow field after velocity propagation.

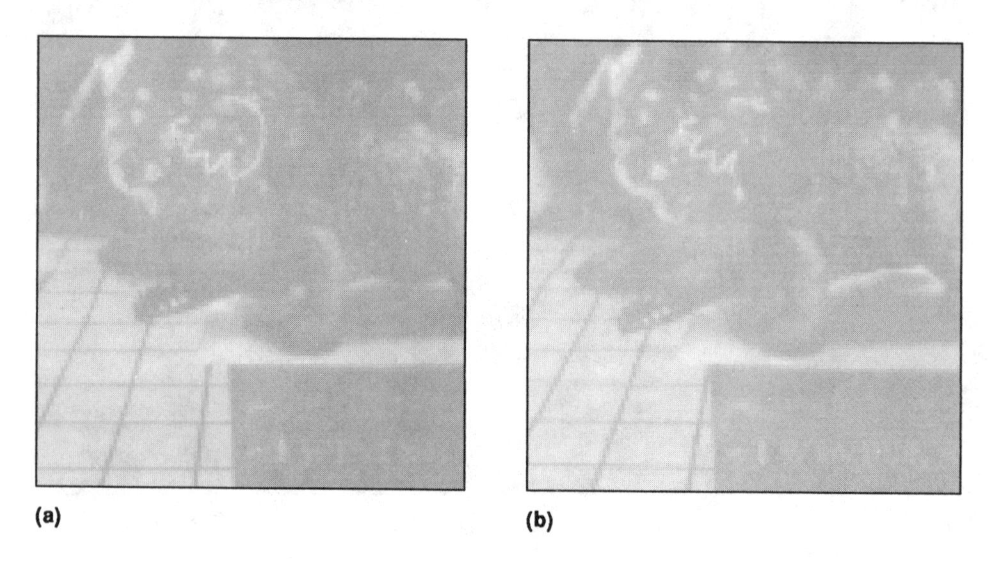

(a) (b)

Figure 6.33. The dinosaur experiment: the two 128×128 images.

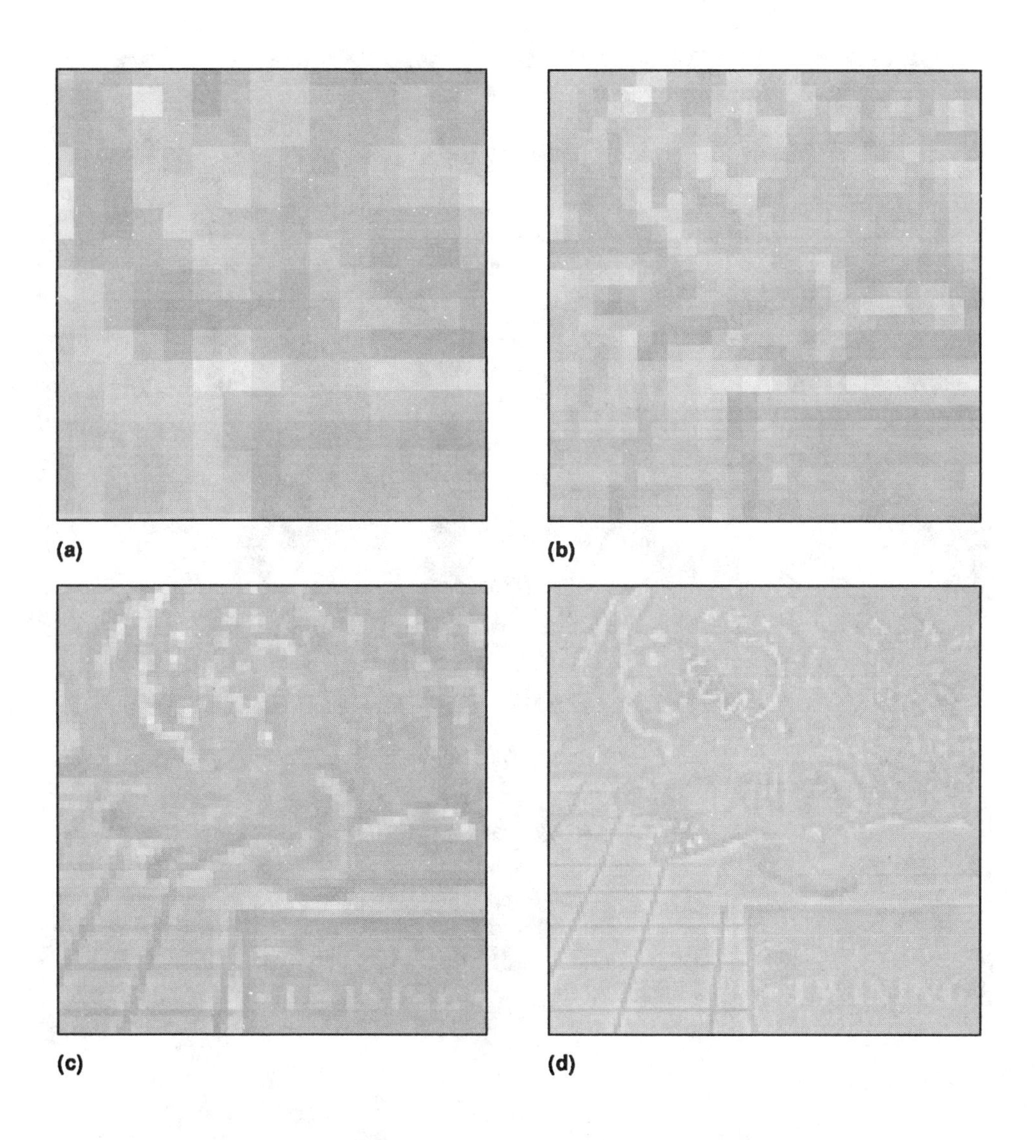

Figure 6.34. The dinosaur experiment: the first frame at (a) level 3, (b) level 2, (c) level 1, and (d) level 0 of the Laplacian pyramid.

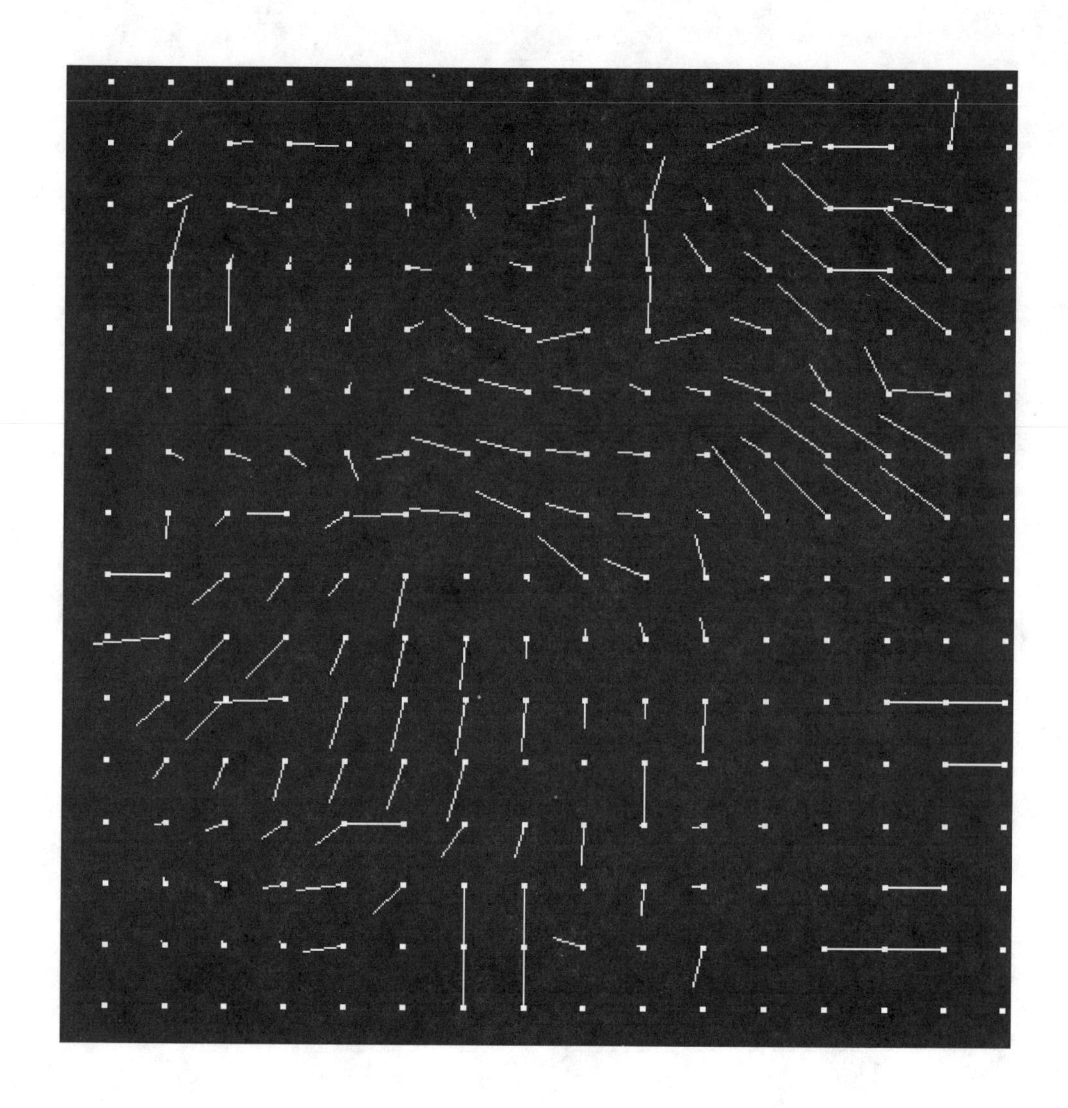

Figure 6.35. The dinosaur experiment: flow field at level 3 before velocity propagation.

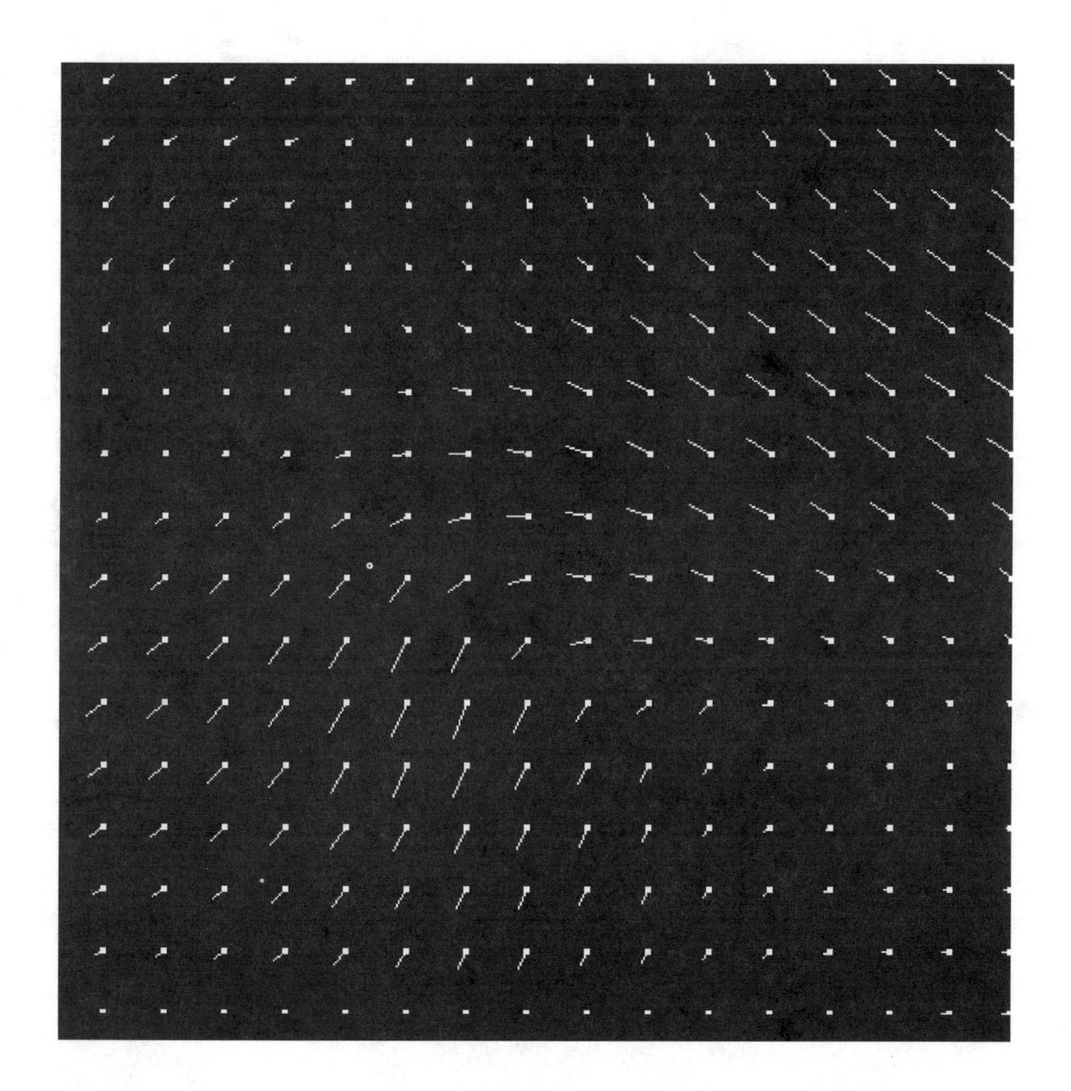

Figure 6.36. The dinosaur experiment: flow field at level 3 after velocity propagation (10 iterations).

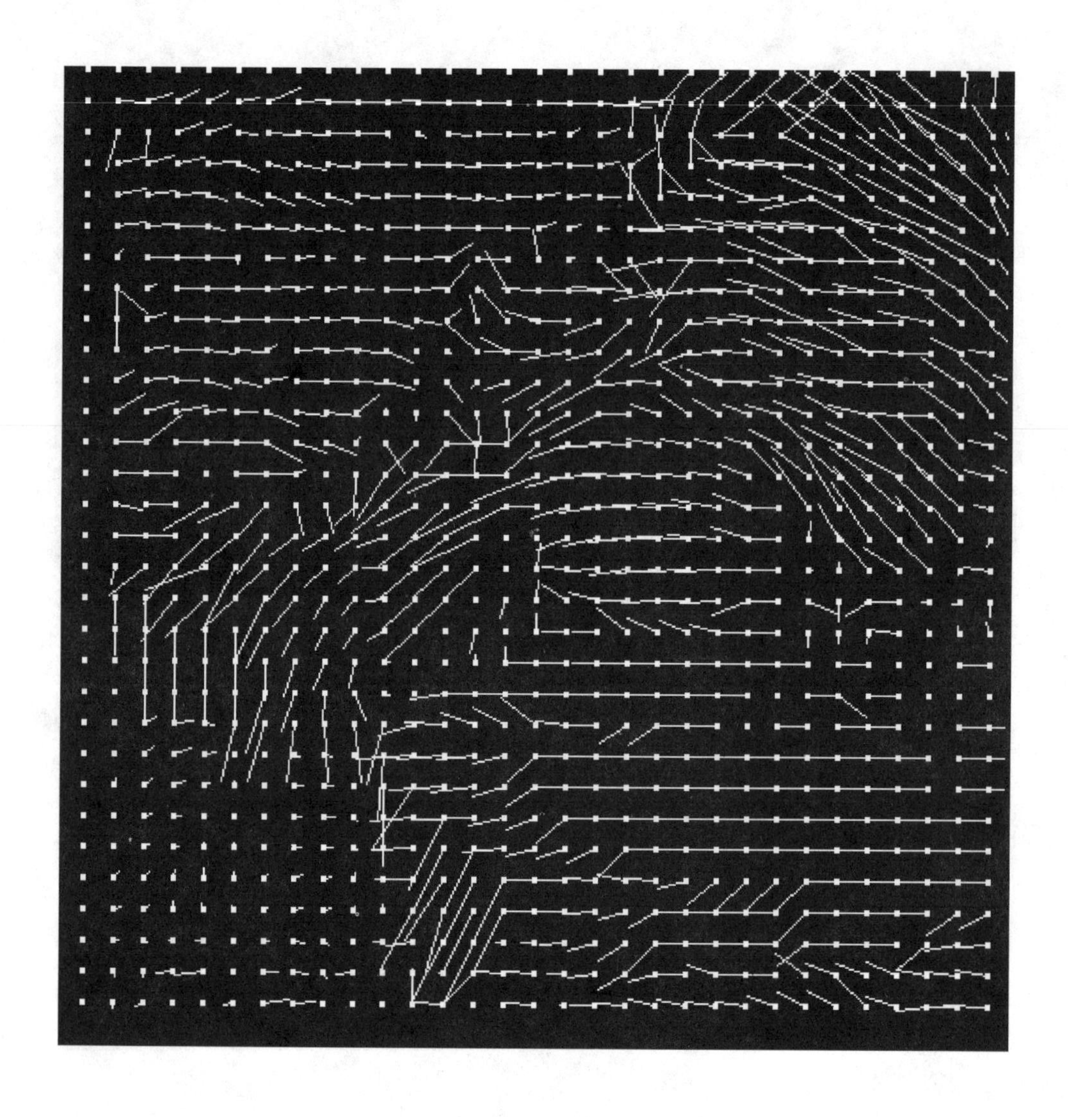

Figure 6.37. The dinosaur experiment: flow field at level 2 before velocity propagation.

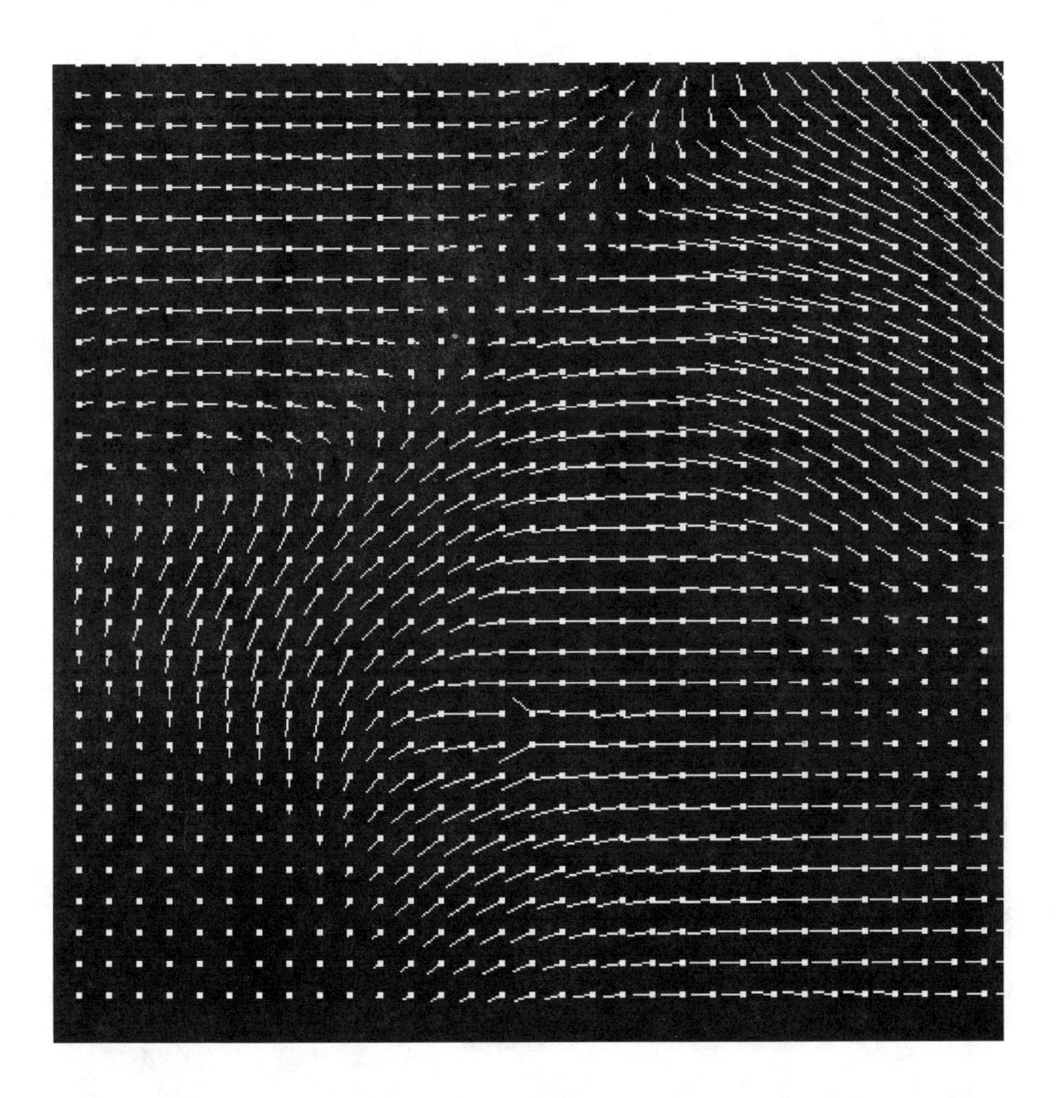

Figure 6.38. The dinosaur experiment: flow field at level 2 after velocity propagation (10 iterations).

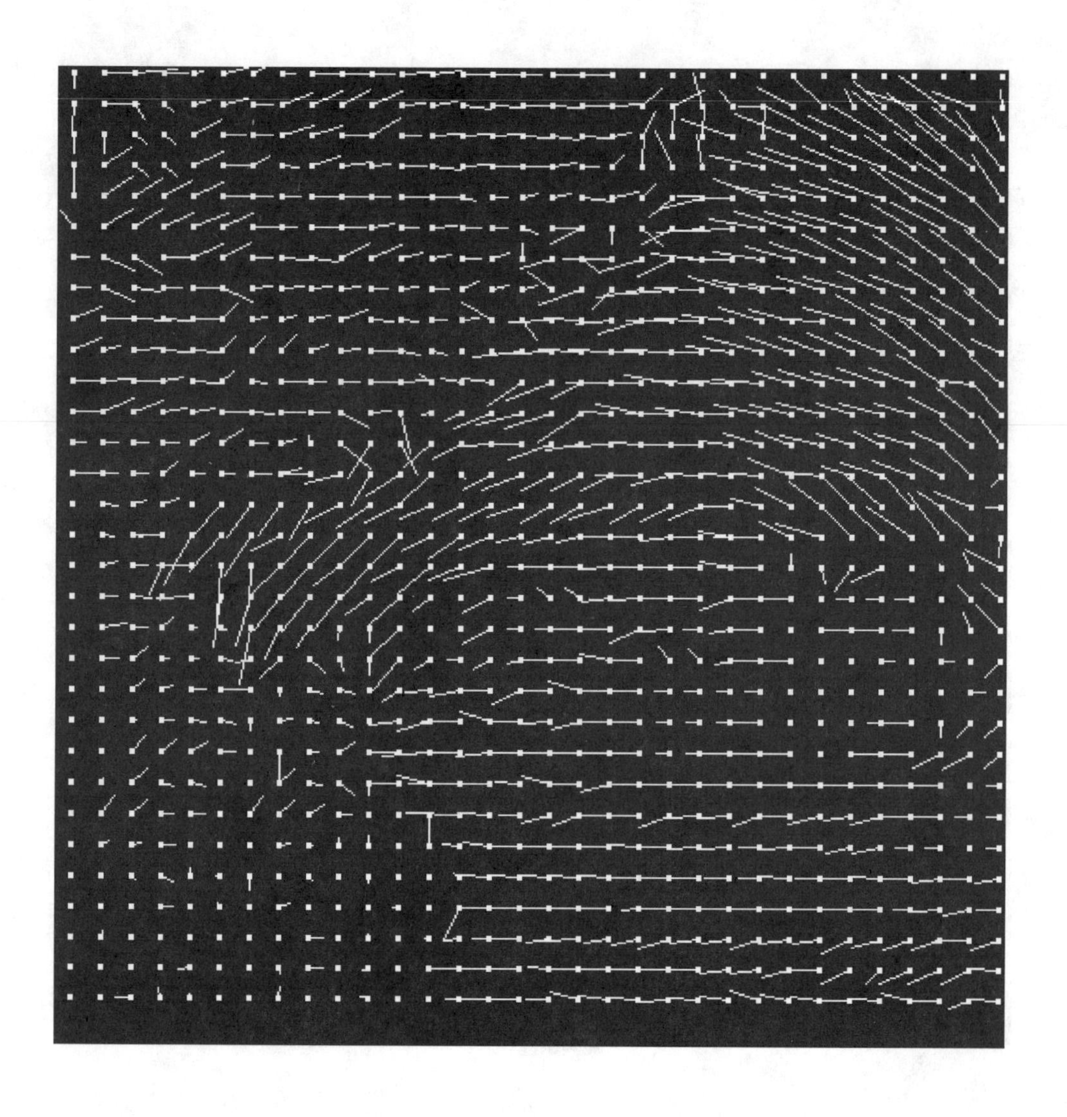

Figure 6.39. The dinosaur experiment: flow field at level 1 before velocity propagation.

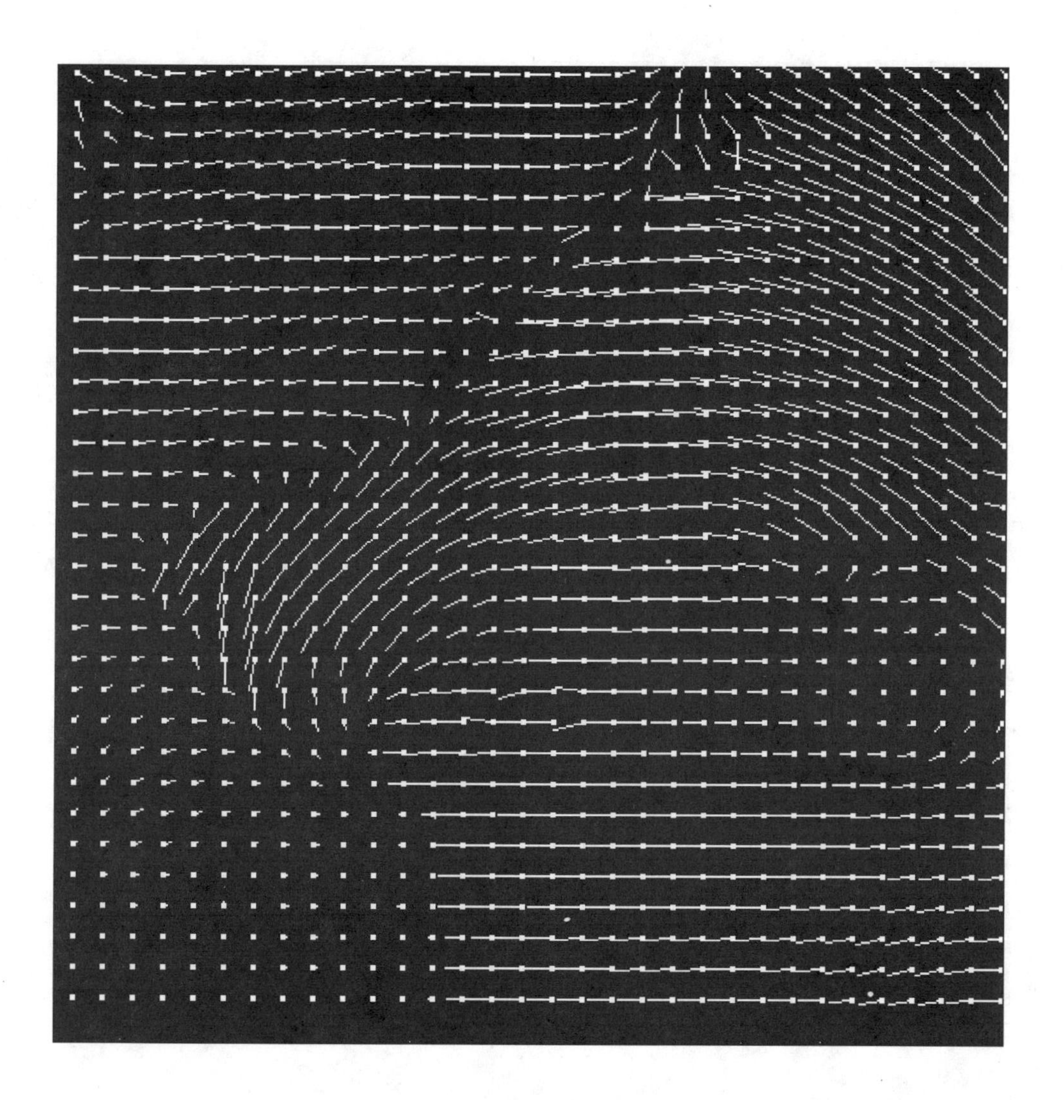

Figure 6.40. The dinosaur experiment: flow field at level 1 after velocity propagation (10 iterations).

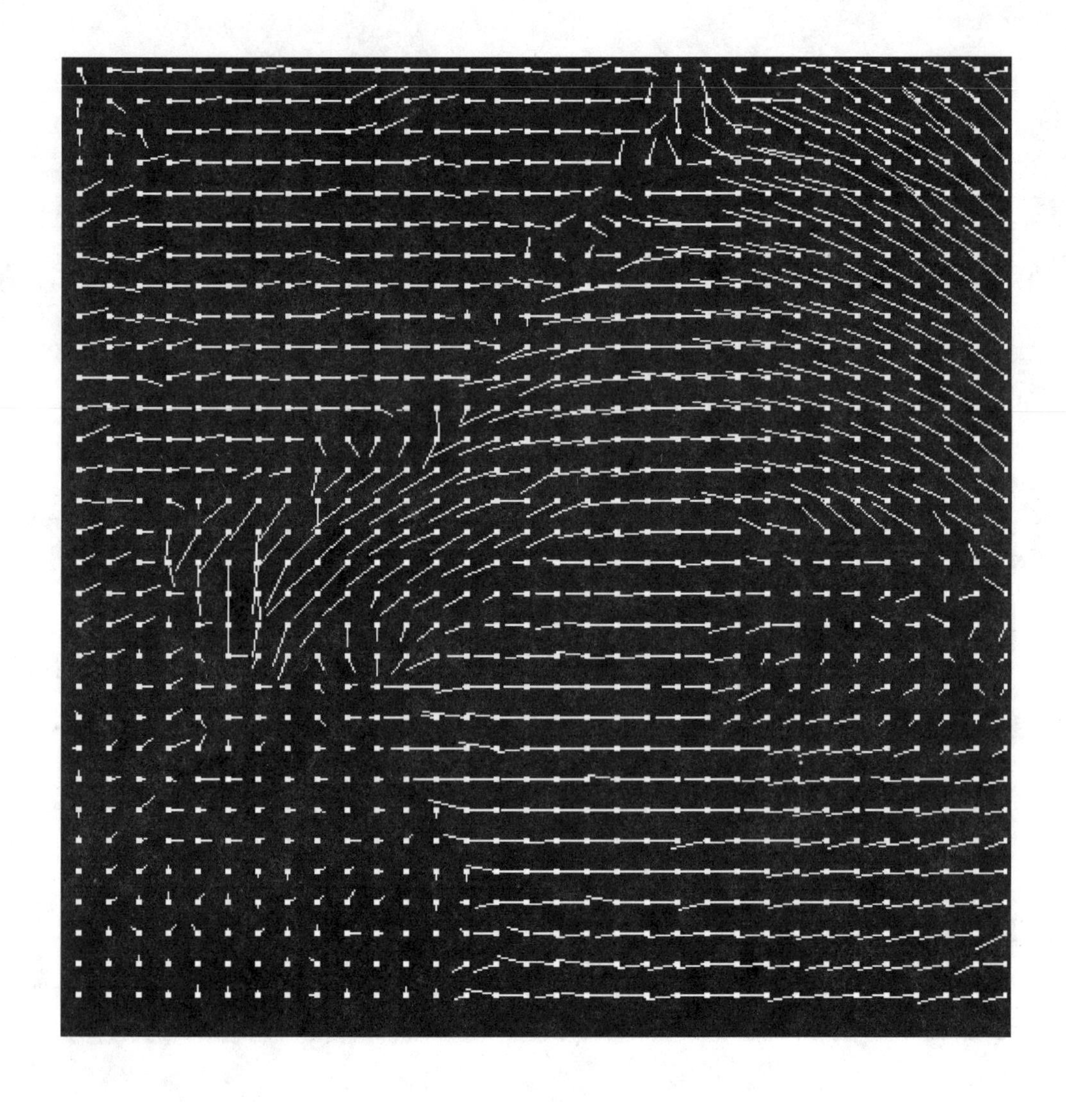

Figure 6.41. The dinosaur experiment: flow field at level 0 before velocity propagation.

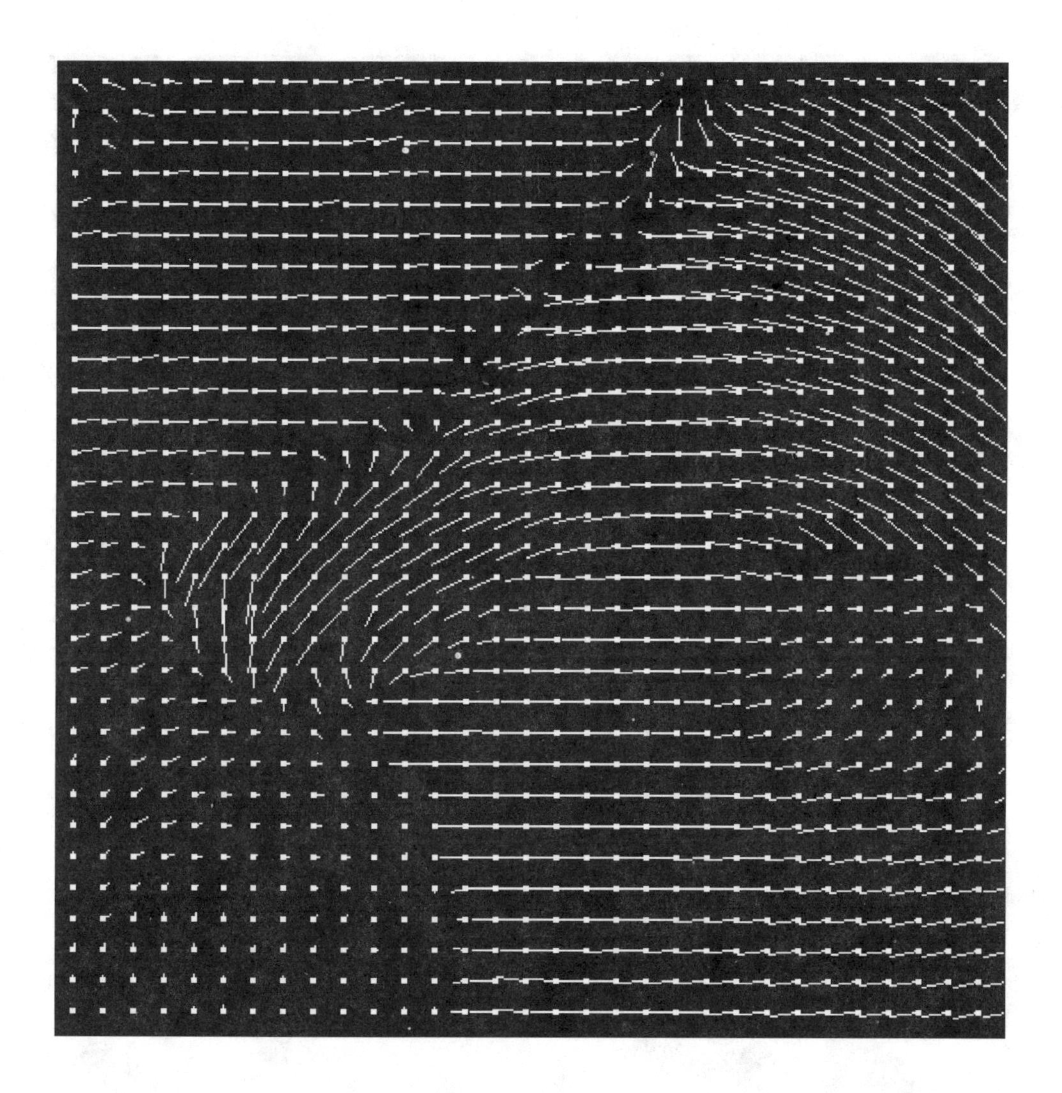

Figure 6.42. The dinosaur experiment: flow field at level 0 after velocity propagation (10 iterations).

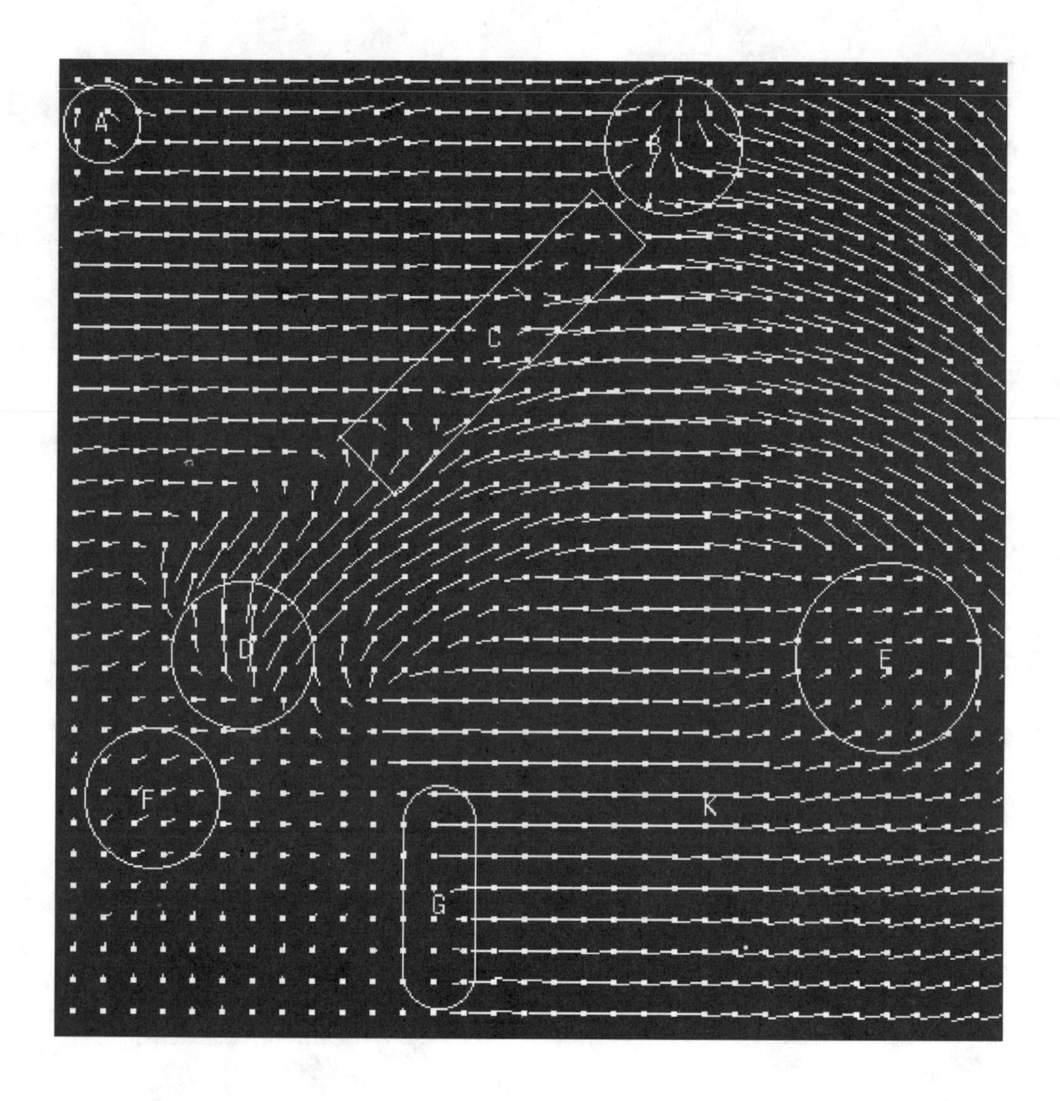

Figure 6.43. The dinosaur experiment: flow field at level 0 after velocity propagation (10 iterations), showing some interesting regions.

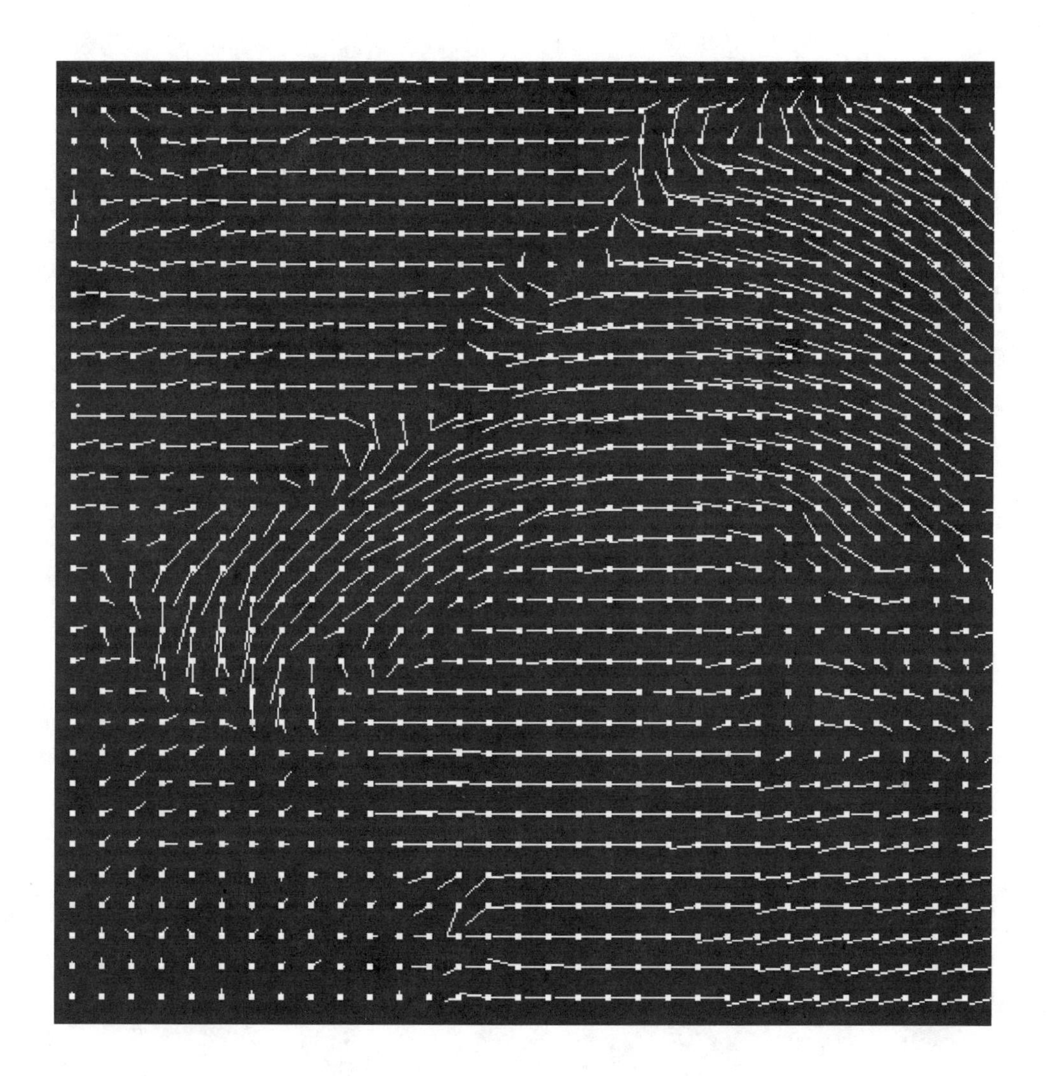

Figure 6.44. The dinosaur experiment: flow field at level 0 after 10 iterations of conventional smoothing.

Chapter 7

Experiments on Application of Optic Flow

The objective of this chapter is to put the framework for optic-flow estimation in the context of some applications. As discussed in Chapter 2, the applications of optic flow can be broadly classified into two categories: three-dimensional and two-dimensional. I will use estimation of scene depth and motion-compensated image-sequence enhancement as applications representing these two categories. After I describe experiments on these two applications, I give some concluding remarks.

Estimation of scene depth

I discussed the basic theory for recovering scene depth from optic flow in Chapter 2. In the recent past, several schemes based on this theory have been proposed. Some of these schemes estimate depth in an on-line, incremental fashion over time, using a Kalman filter. These are of specific interest in this chapter because of the estimation-theoretic nature of the underlying framework for optic-flow computation. One such scheme was demonstrated by Matthies, Szeliski, and Kanade [117]. It requires that an estimate of optic flow be produced along with its covariance for each new frame acquired (in a time sequence) and that they be used to update the existing estimate of disparity (reciprocal of depth) and its variance. Either of the two algorithms discussed in Chapter 5 can be used as a part of this scheme to recover scene depth. Appendix B gives a brief description of the procedure, which is applied to obtain scene depth from optic-flow estimates in the experiments discussed below. For brevity, I present only two experiments: one qualitative and the other quantitative.

To examine the procedure given in Appendix B qualitatively, the toy truck experiment of Chapter 6 was repeated with the truck stationary. The camera looks from the top (about 15 inches above the truck) and undergoes a one-dimensional translation (see Appendix B) in a plane perpendicular to its optical axis. Eleven frames are shot at regular intervals as the camera translates horizontally by 1.5 inches. Figure 7.1 shows the ground-truth depth map (obtained using a laser range finder). Figure 7.2 plots the depth map obtained after 11 frames. For comparison, Figure 7.3 plots the depth map obtained after 11 frames using the optic-flow estimates obtained from the smoothing-based implementation described earlier. It is apparent that the blurring of depth discontinuities is much more prominent in Figure 7.3.

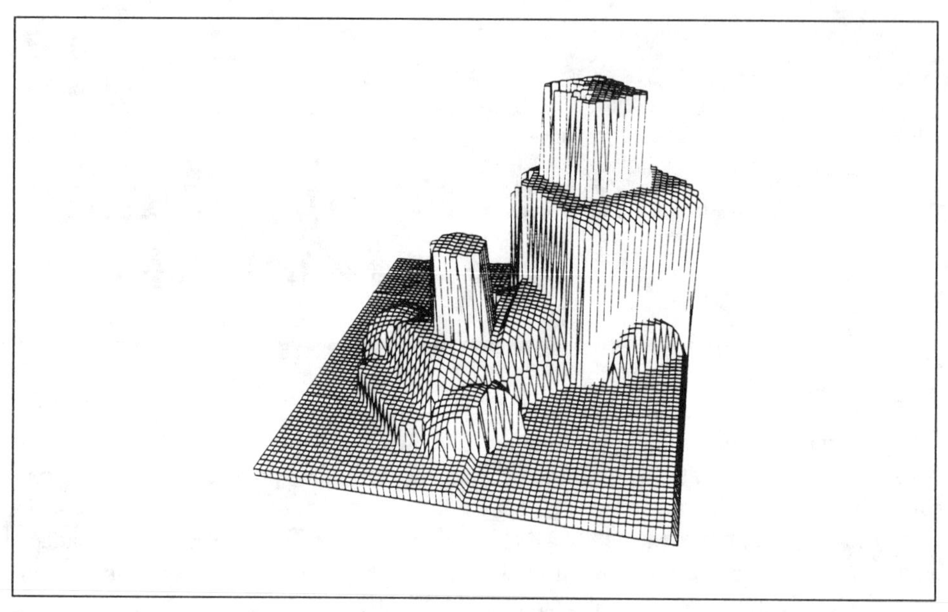

Figure 7.1. The toy truck experiment: a plot of the depth map obtained using a laser range finder.

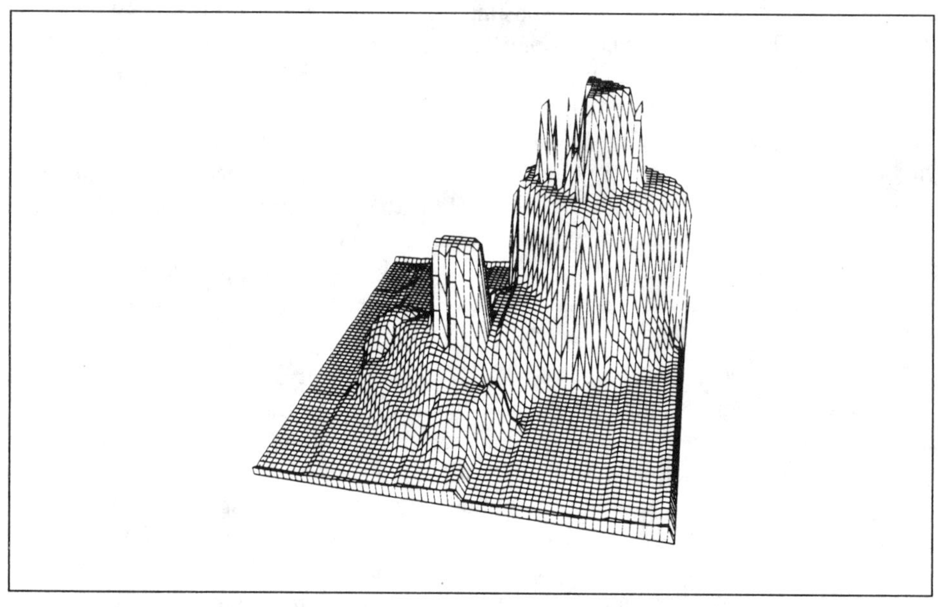

Figure 7.2. The toy truck experiment: a plot of the depth map after 11 frames using estimation-theoretic optic-flow computation.

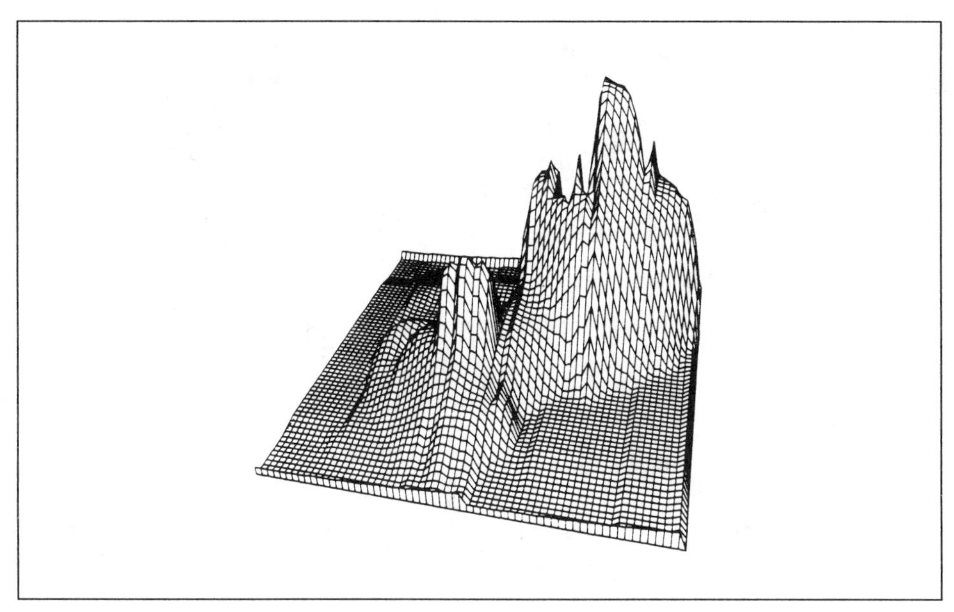

Figure 7.3. The toy truck experiment: a plot of the depth map after 11 frames using conventional smoothing-based optic-flow computation.

Next, a variant of the poster experiment of Chapter 6 was performed to get a quantitative look at the depth estimates. This experiment uses 11 frames, shot at regular intervals as the camera translates horizontally by 0.5 inches, starting from the initial configuration described before, in a plane perpendicular to its optical axis. Figure 7.4 shows the correct depth map. Figures 7.5, 7.6, and 7.7 show the depth map recovered by the procedure after three, seven, and 11 frames, respectively. It is apparent that the depth estimates improve with time. Quantitatively, the root-mean-squared error in depth (over the entire image) is 11.2, 4.3, and 2.8 percent after three, seven, and 11 frames, respectively.

The objective of this exercise (of depth estimation) is to put the new framework in the context of an application, rather than to make any claims about the performance of a specific depth-estimation scheme. Further, even though the experiments shown perform very well in terms of the quality of depth estimates, such performance is an exception rather than a rule. A 1 percent error in optic flow could produce, for example, a 100 percent error in depth. Careful setting of operating conditions is essential to all the algorithms.

Motion-compensated image-sequence enhancement

As discussed in Chapter 2, motion compensation is commonly used to preserve moving regions during temporal smoothing [123, 124]. The simplest case of motion-compensated enhancement uses two frames. Here, optic flow is computed using the two frames, and the first frame is warped using the flow field. The warped frame and the

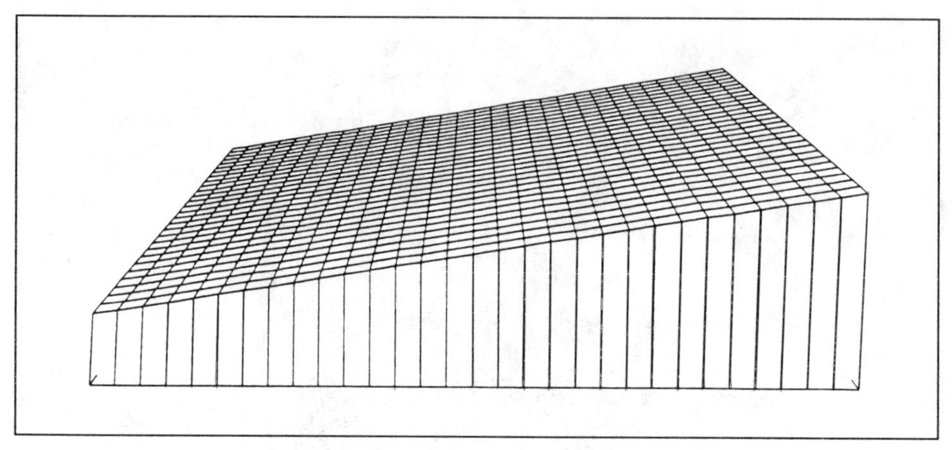

Figure 7.4. The poster experiment: true depth map.

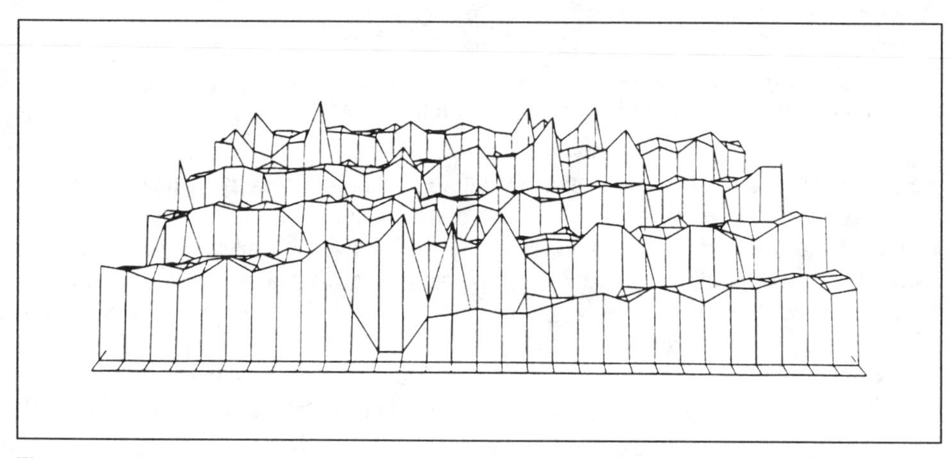

Figure 7.5. The poster experiment: depth map after three frames.

second frame are then averaged to yield the "enhanced" image. Since there is no interframe motion between the two frames, temporal averaging does not introduce any blurring. This simple scheme can be conveniently embedded within a recursive filter, as shown in Figure 7.8.

In this filter, the estimate of intensity $\hat{I}_t$ at a pixel at a given time t is given by

$$\hat{I}_t = \Phi_{t-1} \hat{I}_{t-1} + K(I_t - \Phi_{t-1} \hat{I}_{t-1}) \tag{7.1}$$

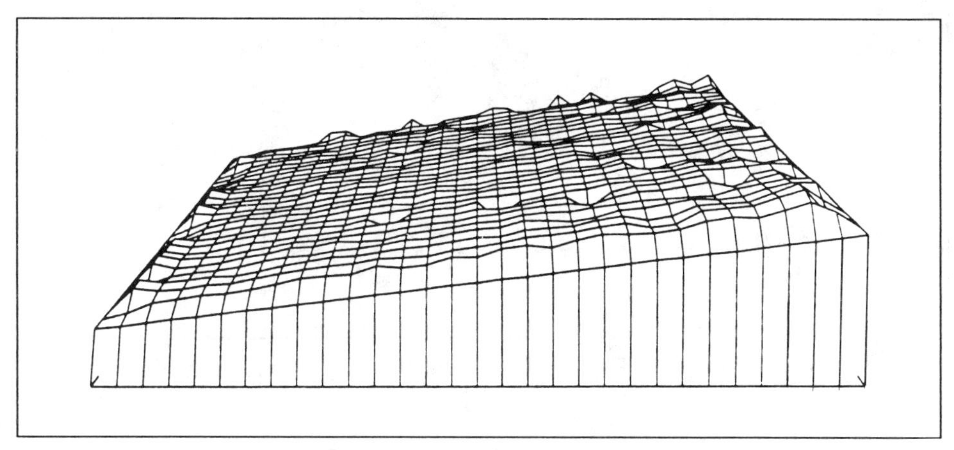

Figure 7.6. The poster experiment: depth map after seven frames.

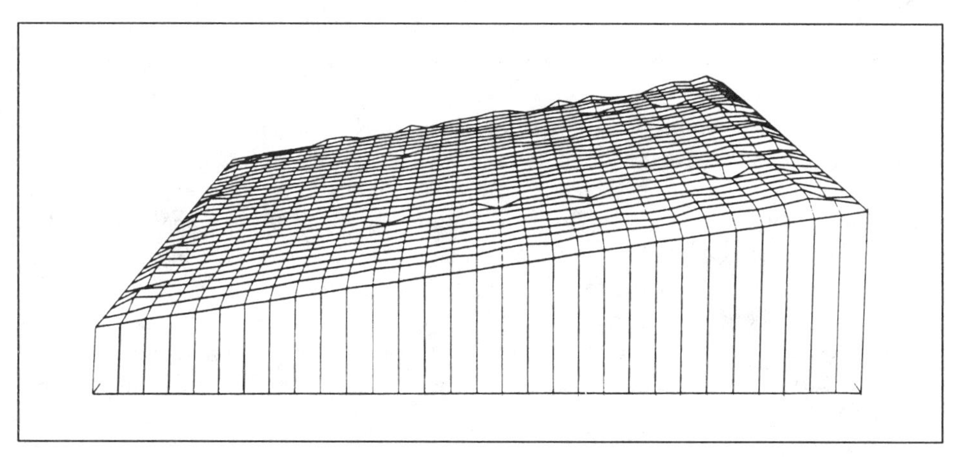

Figure 7.7. The poster experiment: depth map after 11 frames.

where $\Phi_{t-1}\hat{I}_{t-1}$ denotes the intensity at that pixel in the warped version of the estimated image at the previous time instant. Also, K is a gain factor that must be manually assigned. In the following experiments, this factor is chosen to be 0.1. It is apparent that this filter would reduce the noise in an incremental fashion; the images would get better as time progresses. Two experiments based on this scheme are shown below. The first experiment is qualitative and the second is quantitative.

Qualitative experiment. The image sequence used in the qualitative experiment is composed of 15 consecutive frames of an outdoor video movie shot with a *stationary* camera. The images are 256×256 in resolution. The zeroth, second, fourth, sixth,

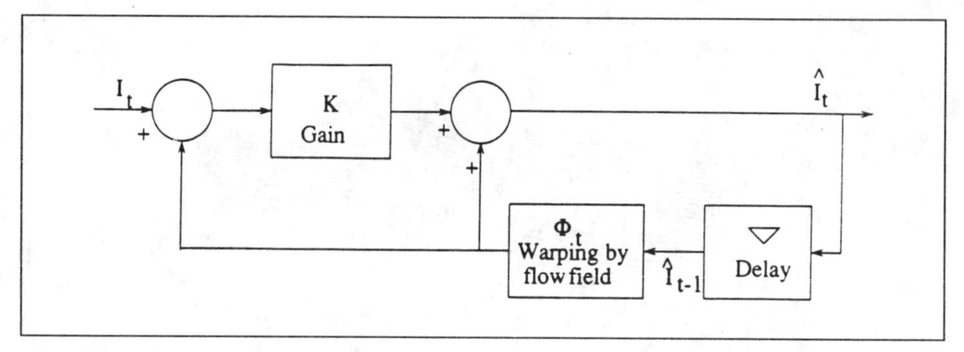

Figure 7.8. A recursive scheme for motion-compensated image-sequence enhancement.

eighth, 10th, 12th, and 14th frames of the original sequence are shown in Figures 7.9a, b, c, d, e, f, g, and h, respectively. It is apparent that there are two moving cars, with stationary buildings, trees, and so on, in the background. The car on the right is occluded by a pole in the foreground.

Zero-mean Gaussian noise with a variance of 20 and 40, respectively, is added to each frame of the original sequence to obtain two noisy sequences (referred to as Sequence 1 and Sequence 2). The zeroth, second, fourth, sixth, eighth, 10th, 12th, and 14th images of the two noisy sequences are shown in Figures 7.10 and 7.11, respectively. The corresponding frames of these two sequences after filtering are shown in Figures 7.12 and 7.13, respectively. It is apparent in Figure 7.12 that the filtered imagery has very little noise after about six frames. In Figure 7.13, on the other hand, a comparable level of noise is achieved after about 12 frames. I do not pursue any quantitative analysis of noise in this experiment because the noise characteristics of the original sequence are unknown.

For comparison, each of the two noisy sequences is filtered using a five-frame mean filter. The eighth frame of each of the two mean-filtered sequences is shown in Figures 7.14a and 7.14b, respectively. It is obvious that there is significant blurring at the regions in motion, that is, the cars. A comparison of Figure 7.12e with Figure 7.14a and of Figure 7.13e with Figure 7.14b reveals that my technique does a very good job of preventing the motion blur.

Quantitative experiment. The image sequence used in the quantitative experiment consists of 15 frames with 128×128 resolution. A dark square, about 40×40 pixels in size, translates downward against a lighter background. The intensity of the background is constant (equal to 128). The intensity of the square increases linearly from zero at its upper edge to 20 at its lower edge. In other words, the vertical gradient of intensity is 0.5/pixel. The horizontal gradient is zero. The motion of the square is one pixel per frame, downward. Figures 7.15a and 7.15b show the first and last frames of the sequence, respectively.

Two sequences are obtained from the ideal sequence described above by adding zero-mean Gaussian noise with a variance of 10 and 30 to each frame of the original sequence. These sequences are Sequence 1 and Sequence 2, respectively. Figures 7.16a and 7.16b show the first and last frames of Sequence 1; Figures 7.16c and 7.16d show those of Sequence 2.

The enhancement algorithm is run on each sequence and the postfiltering noise variance is computed (over the entire image) at each frame. Figure 7.17 graphs the results, which show that the noise variance reduces as more and more frames are available. Further, the higher the noise, the longer it takes for the image quality to reach an acceptable level (i.e., for the postfiltering noise variance to fall below a specified threshold).

Conclusion

The experiments described in the preceding two sections represent an effort to get an overall view of the settings in which the framework for optic-flow computation might be useful. I have discussed one example each of 3D and 2D applications: scene-depth recovery and image-sequence enhancement, respectively.

With respect to scene-depth recovery, this framework lends itself very naturally to incremental estimation using Kalman-filtering-based techniques. The two representative experiments shown in the section on scene-depth estimates illustrate the underlying principle. It is, however, important to understand that recovering scene depth from optic flow is an ill-posed problem. While the two experiments shown in this chapter can recover scene depth very well, their accuracy is attributed to highly controlled experimental conditions (including very accurate knowledge of camera motion). Motion-compensated enhancement of image sequences, on the other hand, is essentially a 2D application where the optic-flow field is used only to warp images. The accuracy of the optic-flow fields recovered by our framework appears sufficient for such a task.

There are many more applications of optic flow, such as tracking and segmentation. An exhaustive evaluation of the new framework with respect to each is beyond the scope of this book.

Figure 7.9. The outdoor scene experiment: (a) frame 0, (b) frame 2, (c) frame 4, (d) frame 6,

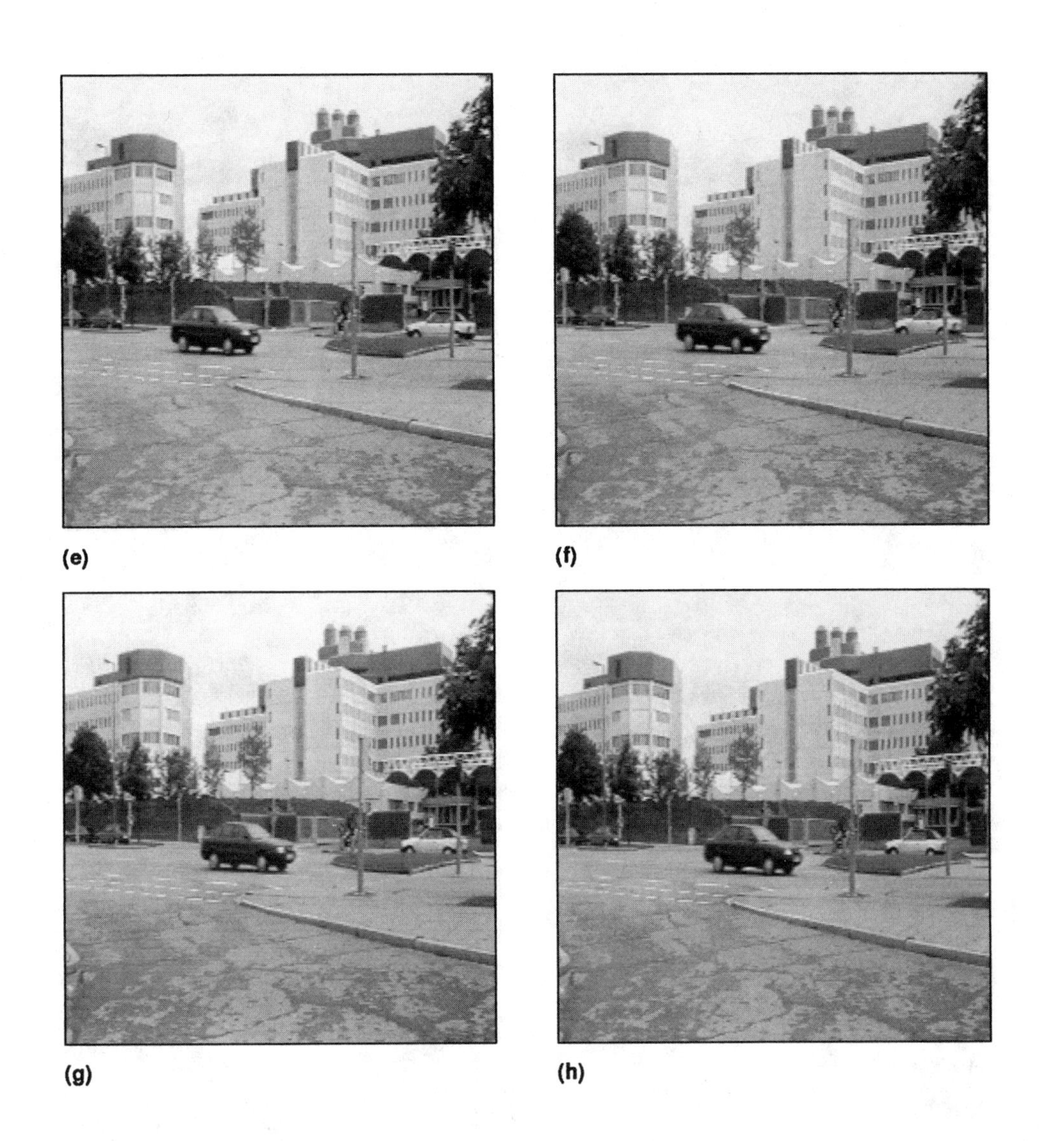

(e) frame 8, (f) frame 10, (g) frame 12, and (h) frame 14 of the original sequence.

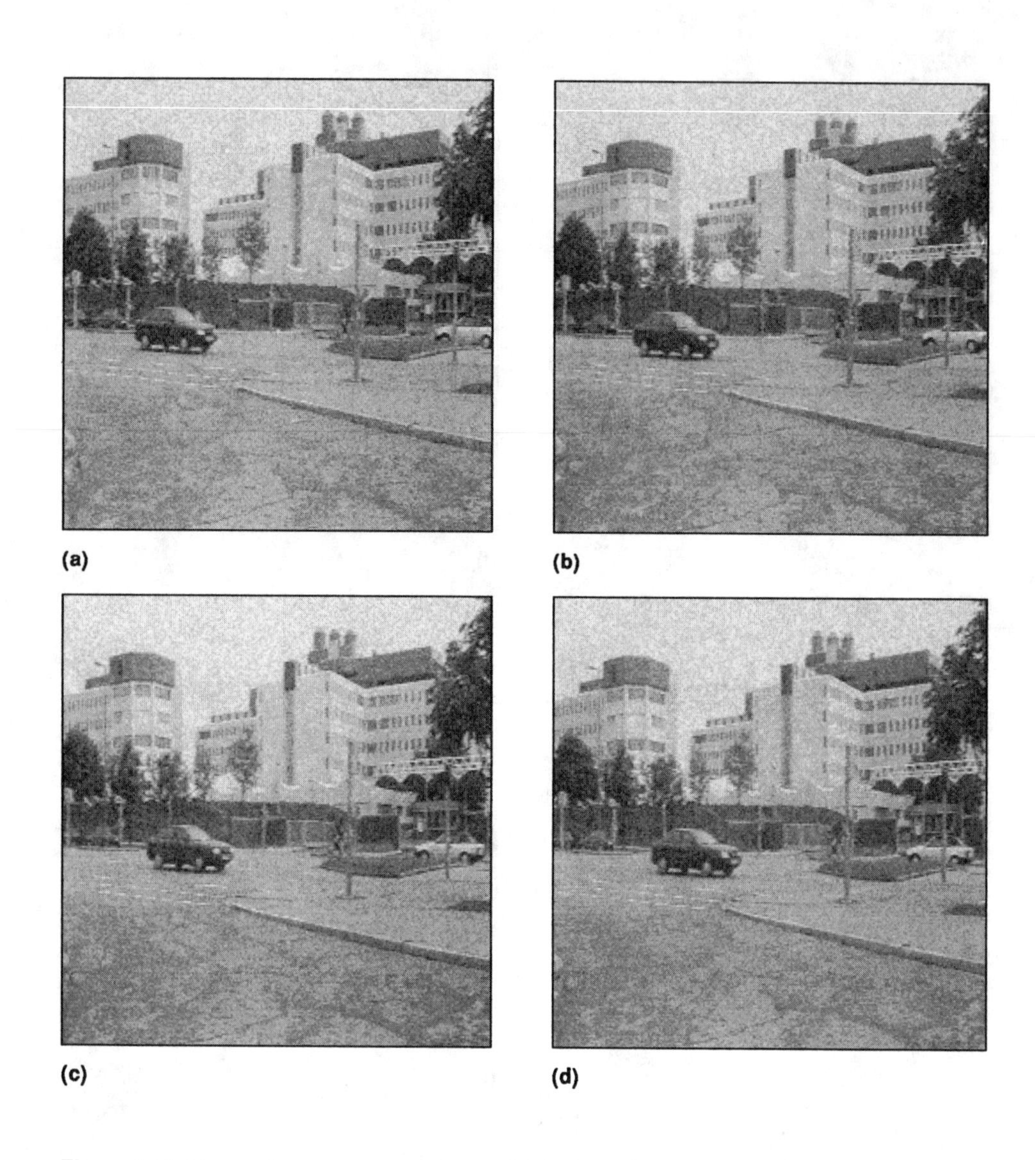

Figure 7.10. The outdoor scene experiment: (a) frame 0, (b) frame 2, (c) frame 4, (d) frame 6,

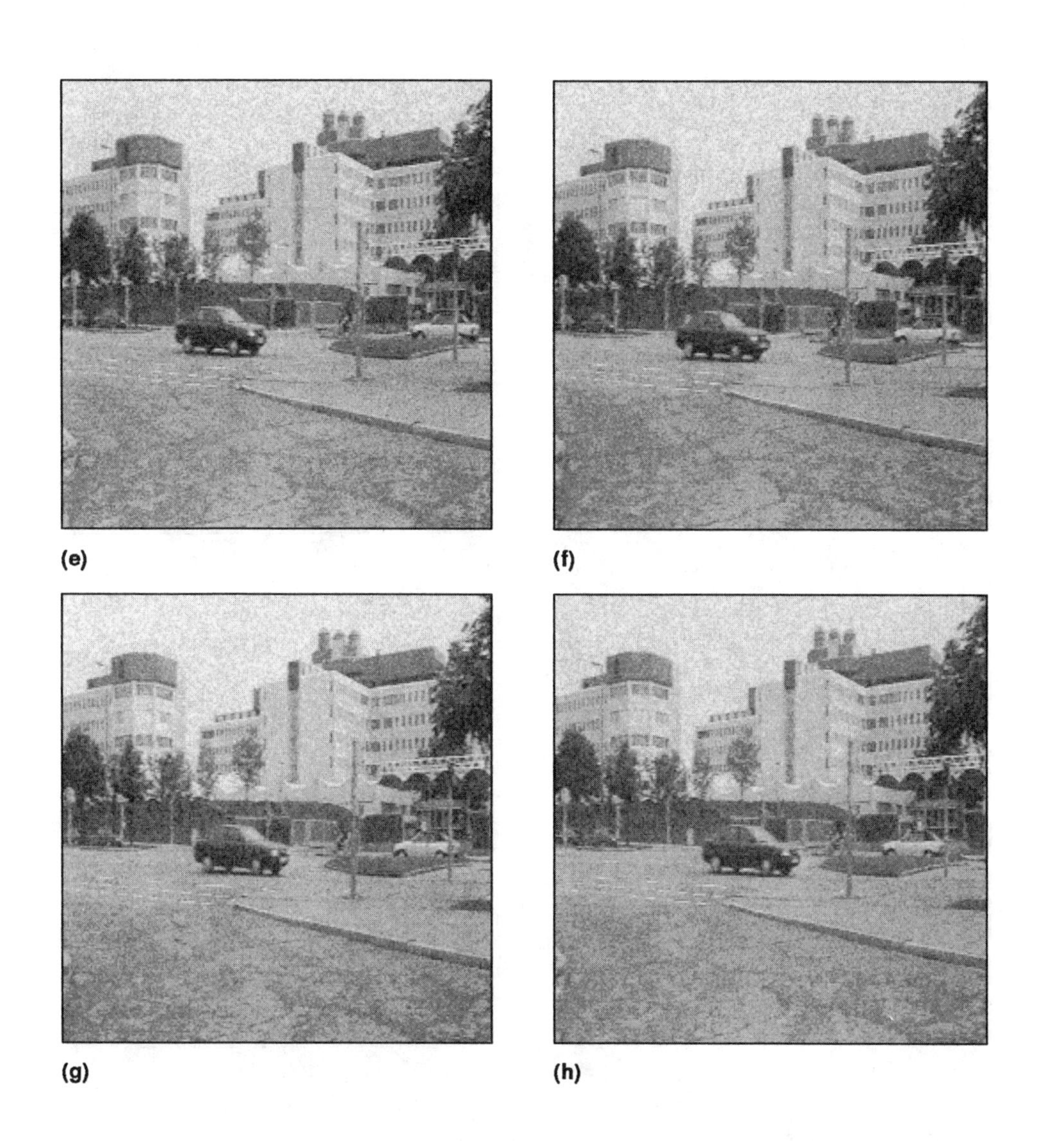

(e) frame 8, (f) frame 10, (g) frame 12, and (h) frame 14 of Sequence 1.

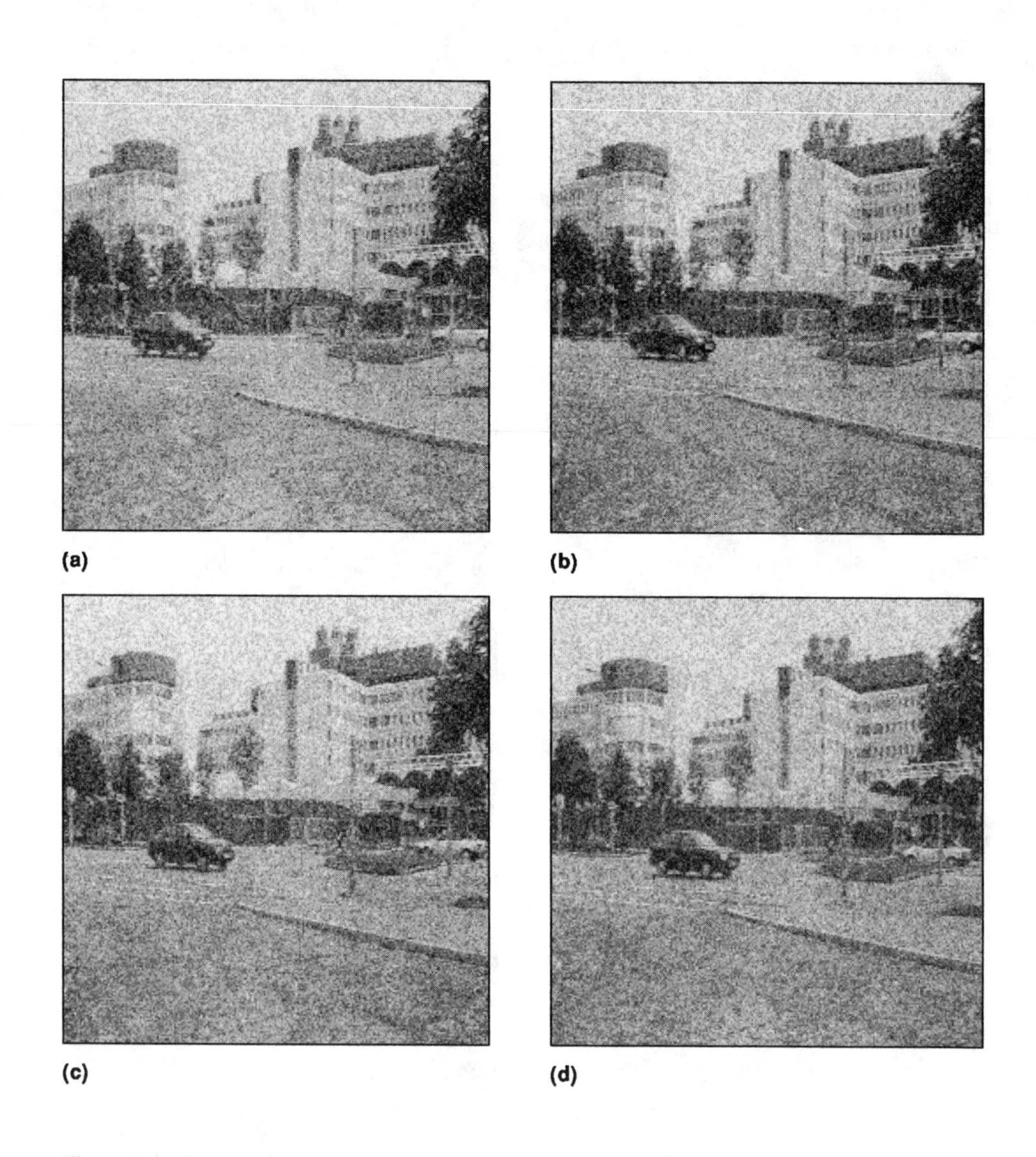

Figure 7.11. The outdoor scene experiment: (a) frame 0, (b) frame 2, (c) frame 4, (d) frame 6,

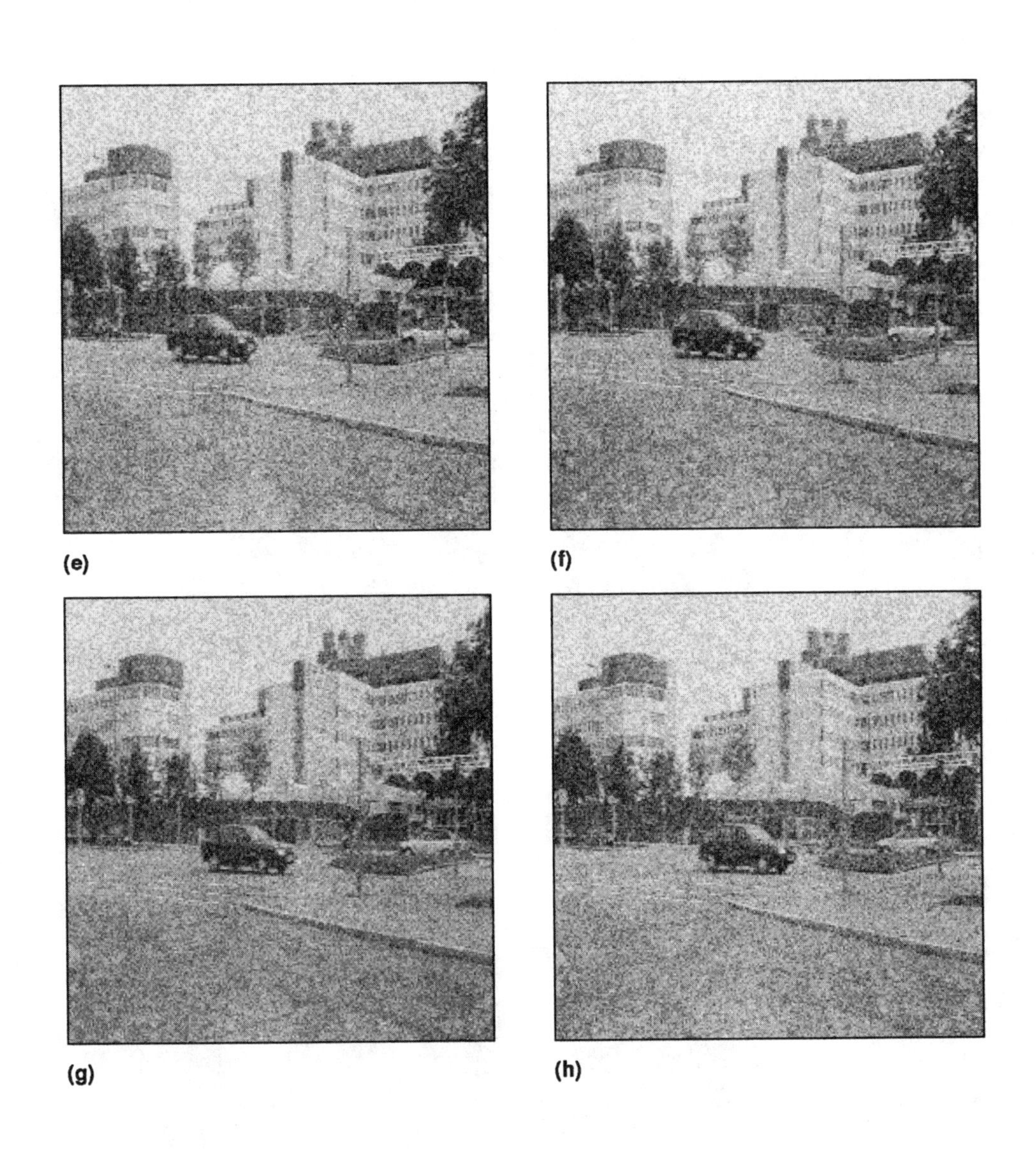

(e) frame 8, (f) frame 10, (g) frame 12, and (h) frame 14 of Sequence 2.

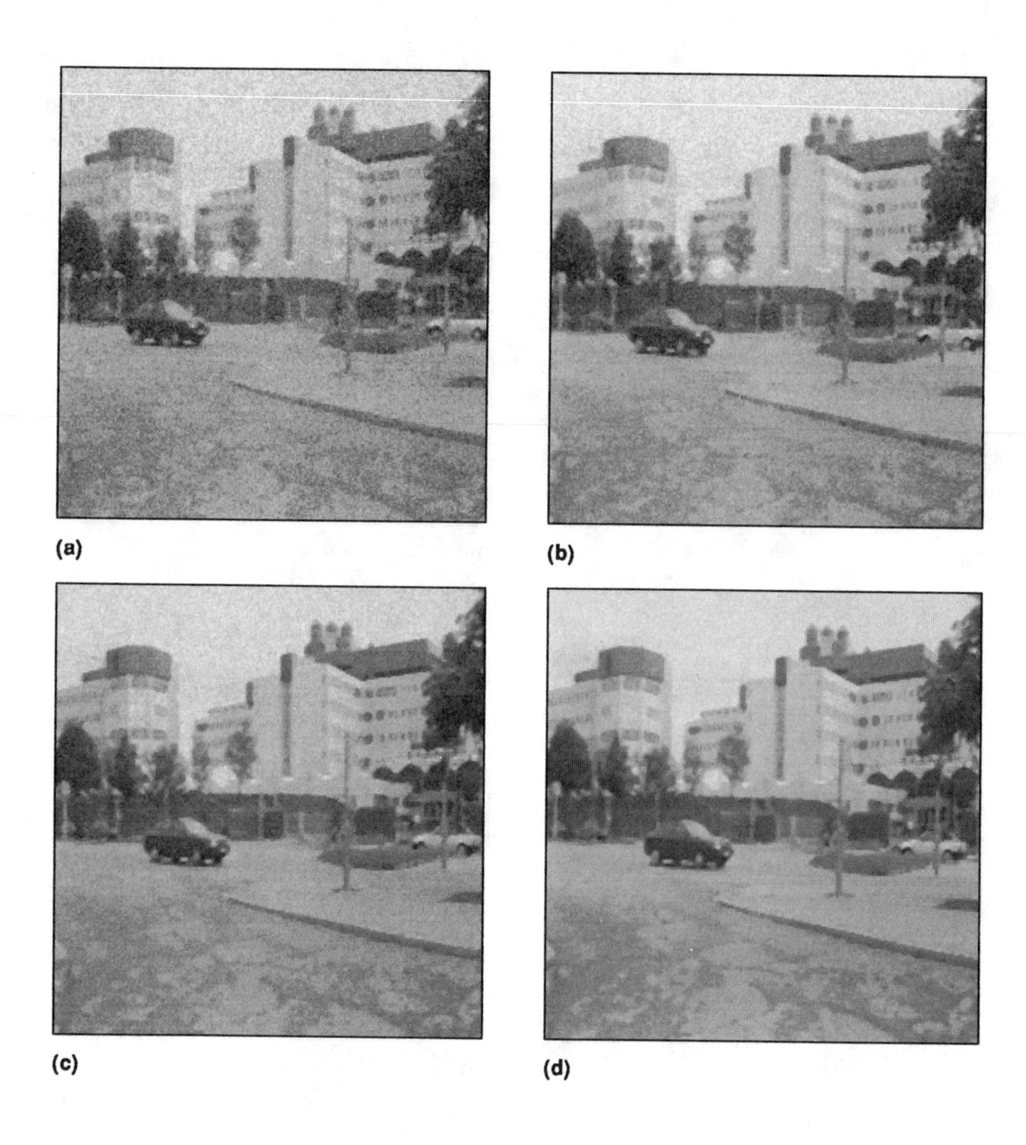

Figure 7.12. The outdoor scene experiment: (a) frame 0, (b) frame 2, (c) frame 4, (d) frame 6,

(e) (f) (g) (h)

(e) frame 8, (f) frame 10, (g) frame 12, and (h) frame 14 of Sequence 1 after filtering.

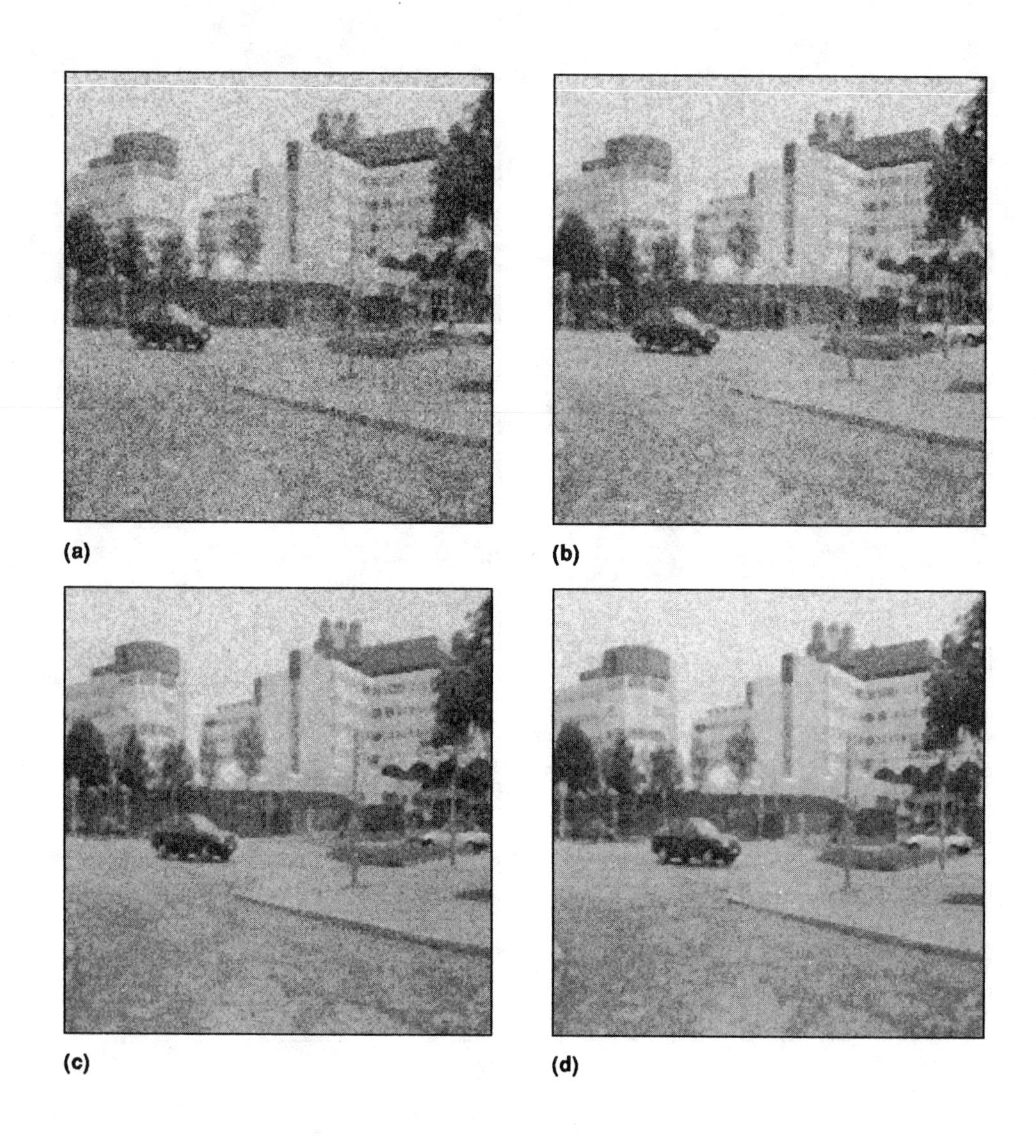

Figure 7.13. The outdoor scene experiment: (a) frame 0, (b) frame 2, (c) frame 4, (d) frame 6,

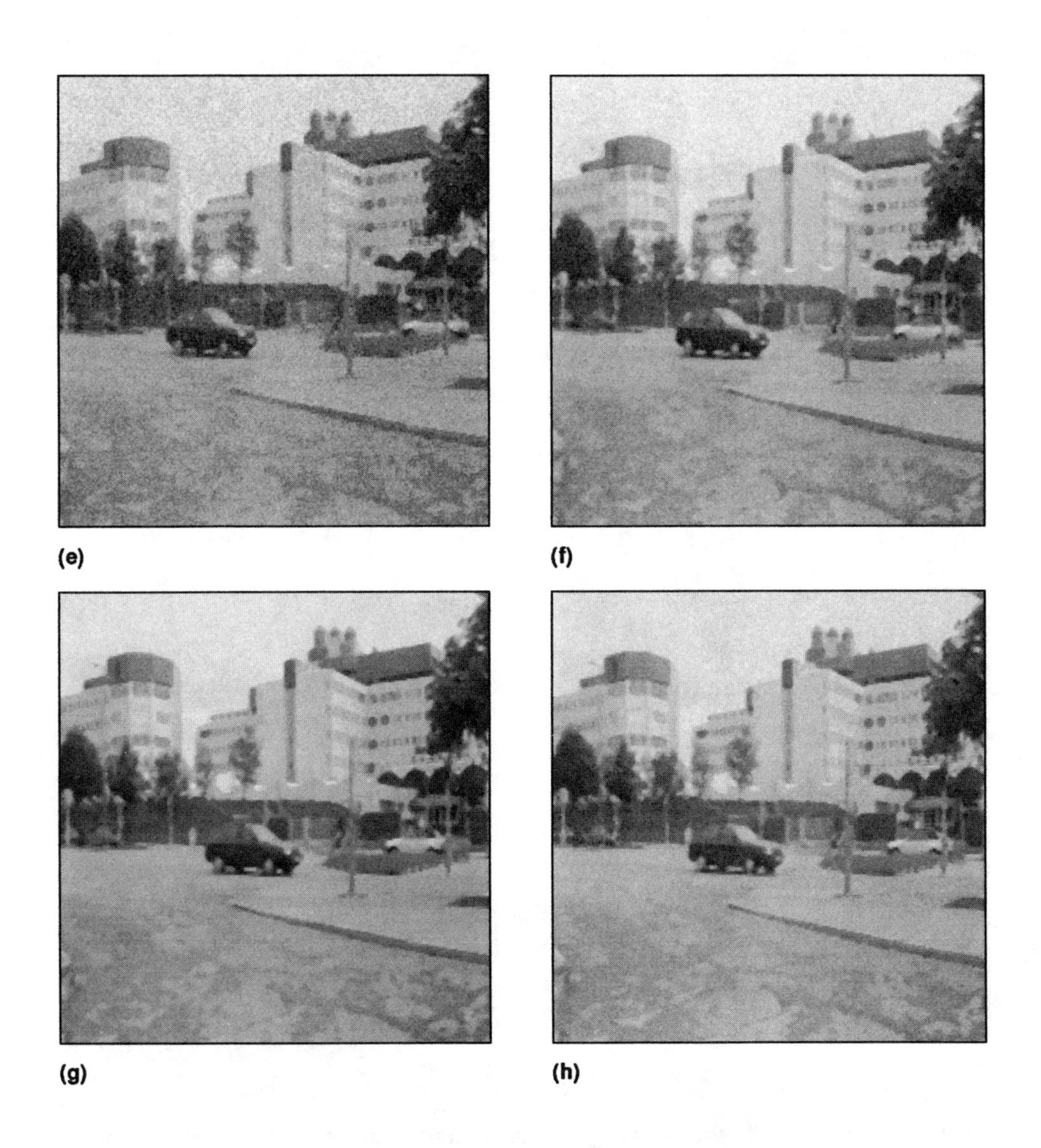

(e) frame 8, (f) frame 10, (g) frame 12, and (h) frame 14 of Sequence 2 after filtering.

Figure 7.14. The outdoor scene experiment: The eighth frame of mean-filtered (a) Sequence 1, (b) Sequence 2.

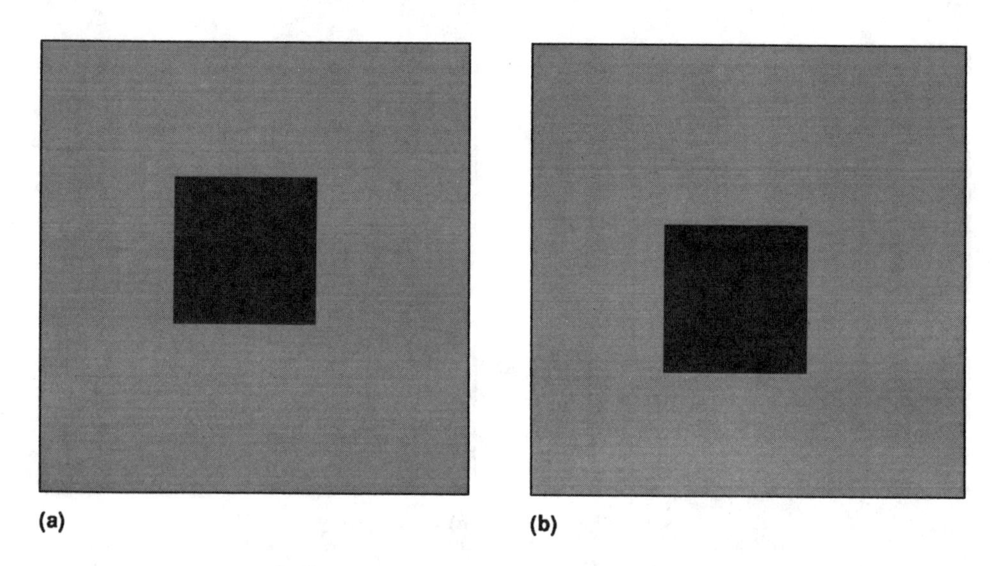

Figure 7.15. The synthetic sequence experiment: the first (a) and last (b) frames of the original sequence.

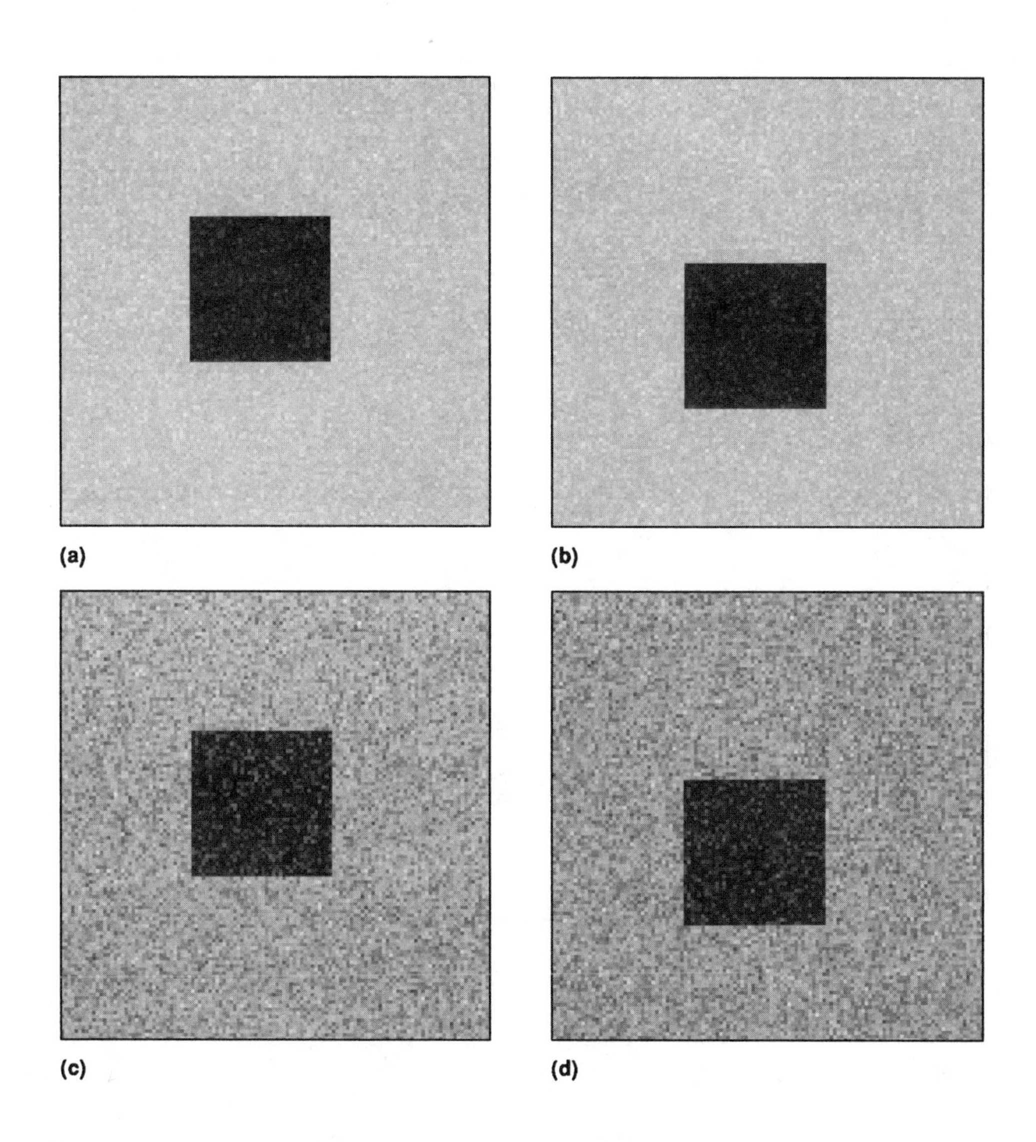

Figure 7.16. The synthetic sequence experiment: the first (a) and last (b) frames of Sequence 1; the first (c) and last (d) frames of Sequence 2.

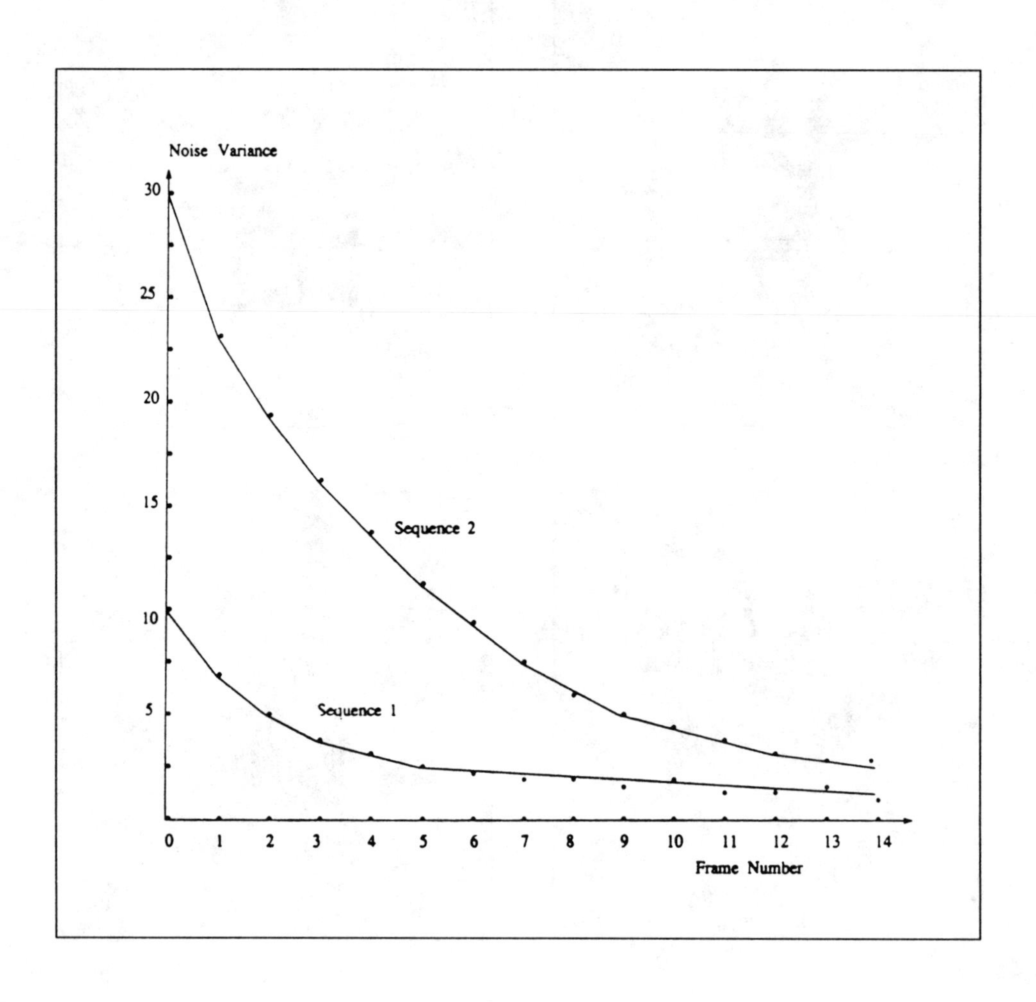

Figure 7.17. The synthetic sequence experiment: Postfiltering noise variance as a function of time (over the entire image).

Chapter 8

Unification

In this chapter, I will show that the framework developed in Chapter 5 serves as a platform to unify various existing approaches for optic-flow estimation. To describe what I mean by unification in this context, I use the expression for the sum of conservation error and neighborhood error. This expression (reproduced from Equation 5.11 with a slight variation of symbols) is

$$\iint [(U - U_c)^T S_c^{-1} (U - U_c) + (U - \bar{U})^T S_n^{-1} (U - \bar{U})] \, dxdy \tag{8.1}$$

where U is the correct velocity at the pixel under consideration, the estimate U_c and the covariance matrix S_c refer to conservation information, and the estimate $\bar{U}$ and the covariance matrix S_n refer to neighborhood information.

In the context of this book, unification has two aspects—one pertaining to the conservation term in the expression given above, and the other pertaining to the neighborhood term. They are discussed below.

In Chapter 5, I used the correlation-based approach to extract conservation information in the requisite form—an estimate accompanied by a covariance matrix. In this chapter, I show that conservation information can also be recovered using the gradient-based and spatiotemporal energy-based approaches—in exactly the same form and using exactly the same steps. In other words, the estimation-theoretic framework of Chapter 5 acts as a unifying framework for the three basic approaches.

In Equation 8.1, the estimate and the covariance matrix representing conservation information are denoted by U_c and S_c, respectively. To indicate the approach used to derive this information, I use the symbols U_{cc} and S_{cc} for the correlation-based approach, U_{cg} and S_{cg} for the gradient-based approach, and U_{ce} and S_{ce} for the energy-based approach.

With respect to neighborhood information, I will show that the approach used in this framework is general and can be reduced to conventional smoothing [34, 80] by replacing the covariance matrix S_n in Equation 8.1 with an approximation valid under specific conditions. In other words, I will show that the framework shown here unifies some of the existing approaches that use conventional smoothing for velocity propagation.

The first parts of this chapter are devoted to conservation information: how to obtain it using the gradient-based and spatiotemporal energy-based approaches. Various procedures developed in each section are illustrated by simple examples using data obtained from real imagery. The objective of these examples is to illustrate the principle and not to make any claims about the quality of optic flow recovered by the procedure. In this book, I do not pursue development and testing of complete algorithms based on the gradient-based and energy-based approaches, as I did for the correlation-based approach. After presenting these examples, I discuss neighborhood information.

Conservation information: Gradient-based approach

Like the discussion of the correlation-based approach, this presentation has four parts. First, I give a brief qualitative description of the overall procedure. Then, I present the quantitative details. I also discuss various implementation issues. Finally, I show the relationship of this procedure with some existing procedures.

A qualitative description. The basic principle of velocity estimation using the gradient-based approach was discussed in Chapter 3. However, an explicit notion of the reliability of the estimates was not developed. In the following discussion, I present a graphical interpretation of the principle discussed in Chapter 3 to derive a qualitative description of reliability.

To keep the diagrams simple, I use a one-dimensional first-order model. By one-dimensional, I mean there is only one spatial direction, say x. Further, a first-order model implies that the local intensity variation around a point can be considered linear in the spatial coordinates. Imagine a one-dimensional pattern with a ramp intensity profile translating with a uniform velocity to the right. Under the assumption of conservation of intensity, Figures 8.1a and 8.1b show the "images" at two closely spaced time instants. The objective is to measure the displacement of a given point P from one image to the next. Without loss of generality, we can assume that the time elapsed between two successive images is unity. Thus, the displacement is equal to the velocity v. The displacement is equal to the base AB of the triangle ABC, as shown in Figure 8.1c. The measurements that can be made on the imagery are

- the temporal derivative of intensity I_t at the point P, which is given by the height BC of the triangle ABC, and
- the spatial derivative of intensity I_x at the point P, which is given by the tangent of the angle θ.

We can compute the displacement AB as the quotient of the two measurements BC and $\tan(\theta)$. As Figure 8.1d shows, for a given value of temporal derivative I_t and for a given probable error $\tan(\theta)$ in the measurements of the spatial derivative, the absolute uncertainty in the estimate of displacement will be *lower* for *higher* values of the spatial derivative $\tan(\theta)$. In other words, the higher the spatial derivative of intensity, the higher the confidence measure associated with the estimate of velocity.

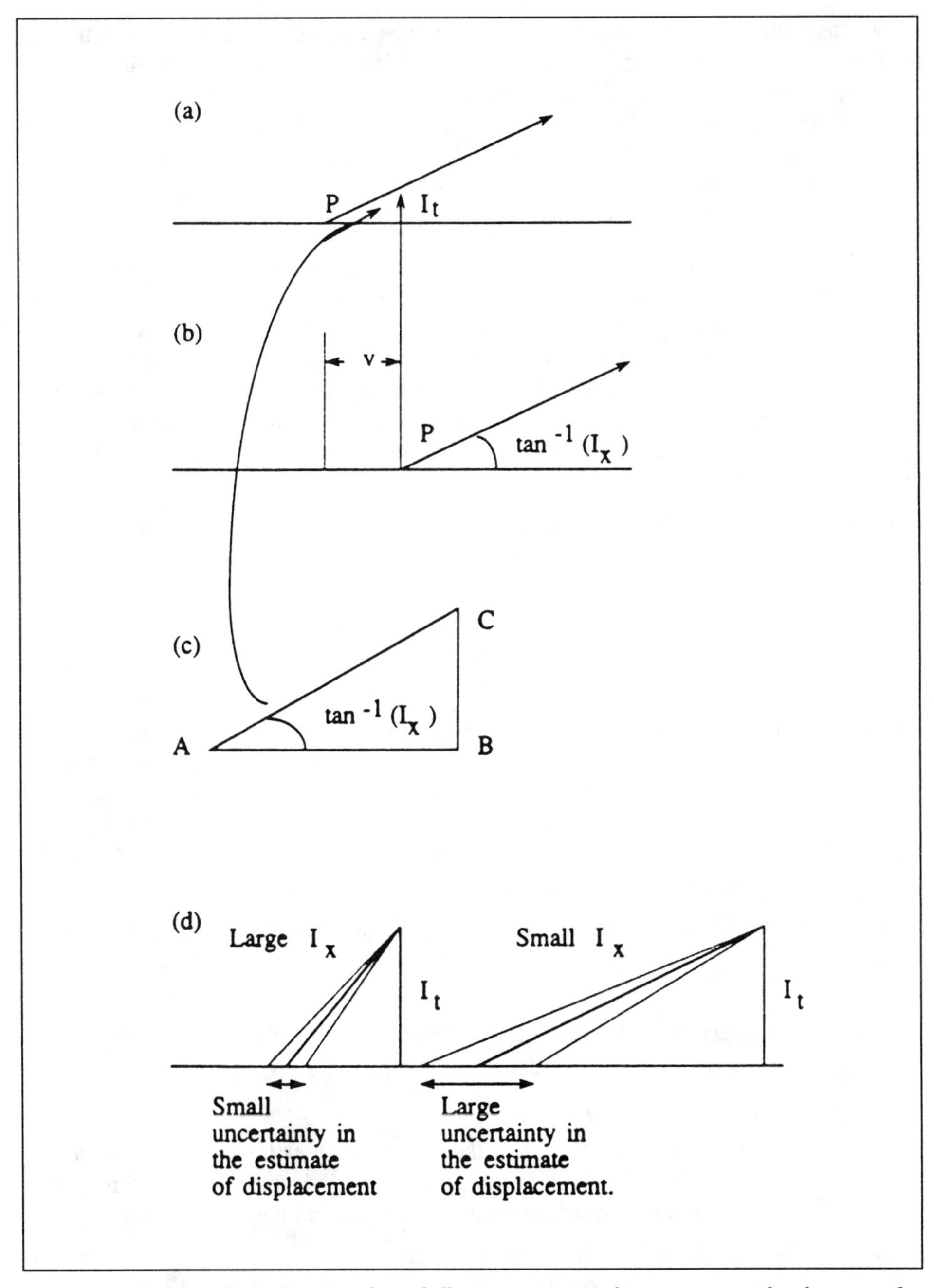

Figure 8.1. Gradient-based estimation of displacement: (a, b) two successive images of a one-dimensional pattern with a ramp intensity profile, (c) estimation of displacement, (d) uncertainty in the estimate of displacement.

By constructing an error distribution in velocity space, we can capture the notion of confidence measure described above. Imagine a point on the moving one-dimensional intensity pattern in Figure 8.1. Assume that the spatial derivative at the point at a given time is I_x and the velocity at the point is v. One time unit later, the *predicted* temporal change in intensity at that point will be given by $-I_x v$ (see Figure 8.1c). Let the *observed* temporal change at the point be given by I_t. Under ideal conditions, I_t should be equal to $-I_x v$. However, because of noise and digitization effects, the two will not be equal.

Let the error between the predicted and the observed values of the temporal change be defined as $(I_x v + I_t)^2$. Assuming that the velocity v is unknown, we can plot the error as a function of velocity (in the vicinity of some a priori expected value) to get an error distribution. Figure 8.2 shows an example. Two patterns with different (known) values of the spatial derivative, moving with a unit of velocity, are considered. The observed values of the temporal derivative are as shown in the figure. Error distribution is plotted in each case. The pit in the error distribution is the most well defined in the case where the spatial derivative is the highest.

As in the case of the correlation-based approach, response distribution can be defined as

$$e^{-k(I_x v + I_t)^2}$$

Figure 8.2 also plots response distribution. As expected, the peak in response distribution is the most well defined in the case where the spatial derivative is the highest. The mean of the response distribution serves as an estimate of velocity. Furthermore, the variance denotes the uncertainty in the estimate. As with the correlation-based approach, the reciprocal of the variance could be used to denote the confidence measure. The confidence measure increases with the magnitude of the spatial derivative.

The qualitative description given above demonstrates that there are three essential steps underlying the recovery of conservation information using the gradient-based approach:

1. Select the conserved quantity and derive it from intensity imagery.
2. Compute error distribution and response distribution in velocity space (over the search area).
3. Interpret response distribution (i.e., *either* compute an estimate of velocity along with a covariance matrix *or* perform principal-axis decomposition of response distribution to compute two conservation constraints and their confidences).

I now show the quantitative details of these steps.

The computational framework. The reliability of gradient-based computations depends on the ability to reliably estimate the spatial and temporal gradients of the conserved quantity. This factor affects the choice of conserved quantity, and I discuss

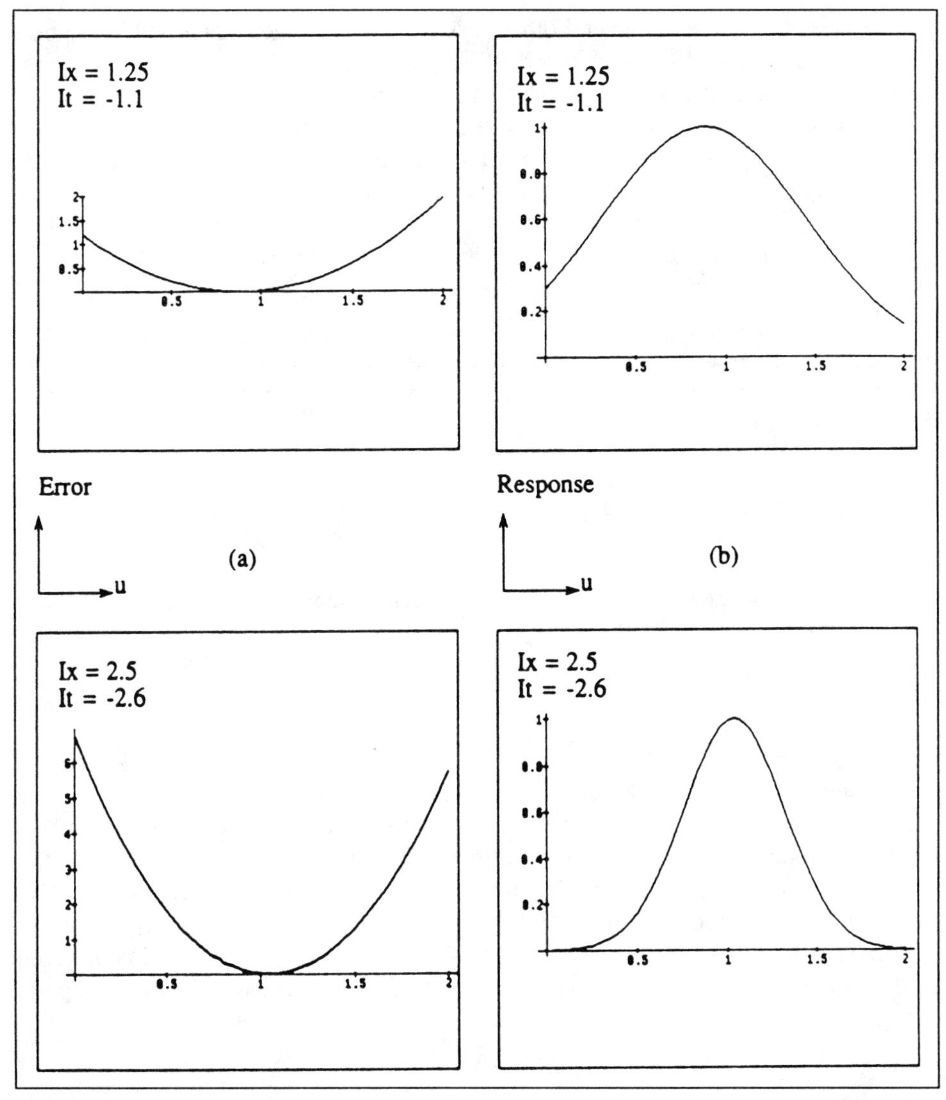

Figure 8.2. Error distribution (a) and response distribution (b) in gradient-based estimation of velocity in the one-dimensional case.

it below. Next, I show the procedure to compute error and response distributions and to interpret response distribution.

Select the conserved quantity and compute it from intensity imagery. The effects of noise and digitization in the original imagery make it difficult to estimate gradients accurately. Hence, image intensity is not a good candidate for the conserved quantity.

As discussed in Chapter 3, several other possibilities have been suggested in the past research. Notable among them are the Laplacian of intensity and spatiotemporal d'Alembertian of intensity [66]. Both have the advantage that they vary almost linearly (with spatial coordinates) in the vicinity of their zero crossings, that is, near intensity edges. If either were used as the conserved quantity, its derivatives could be computed very reliably, but only at the zero crossings. What we need is a quantity in which the derivative can be computed reliably even in (and especially in!) the regions of smoothly varying intensity. Low-pass filtered intensity is a quantity that satisfies this criterion. For this reason, I assume that low-pass filtered intensity is conserved over time.

Under this assumption, if a point P in the scene is in relative motion with respect to the imaging system and it projects onto the locations (x, y) and $(x + \delta x, y + \delta y)$ in the image at time instants t and $t + \delta t$, respectively, then the low-pass filtered intensity I at the two points is equal. That is,

$$I(x, y, t) = I(x + \delta x, y + \delta y, t + \delta t) \tag{8.2}$$

Low-pass filtering can be implemented by convolving the image with a Gaussian. I use the 5×5 mask suggested by Burt [137].

Computing error distribution and response distribution in velocity space. The approach suggested in the qualitative description given earlier can be extended to a 2D case as follows. Once again, assume that the variation of intensity in the local neighborhood of any point is linear in spatial coordinates. If I_x and I_y are the spatial gradients of (low-pass filtered) intensity at a point and (u, v) is the instantaneous velocity vector at that point, then the predicted temporal change in (low-pass filtered) intensity over a unit of time is given by $-(I_x u + I_y v)$. The observed temporal change over this time is given by the temporal derivative I_t. The error between the predicted and the observed temporal change can be defined as

$$\mathcal{E}_g(u, v) = (I_x u + I_y v + I_t)^2 \tag{8.3}$$

Under ideal conditions, the error will be zero at (u, v) corresponding to the actual velocity. Because of noise and digitization effects, however, the error is greater than zero. Figure 8.3a shows the general nature of error distribution in velocity space. It can be shown [153] that with the formulation of error above, the iso-error contours are straight lines. Thus, there is a line—not a point—in uv space along which the error is minimal.

I define response distribution using an exponential function (for the same reasons as those explained in the case of the correlation-based approach):

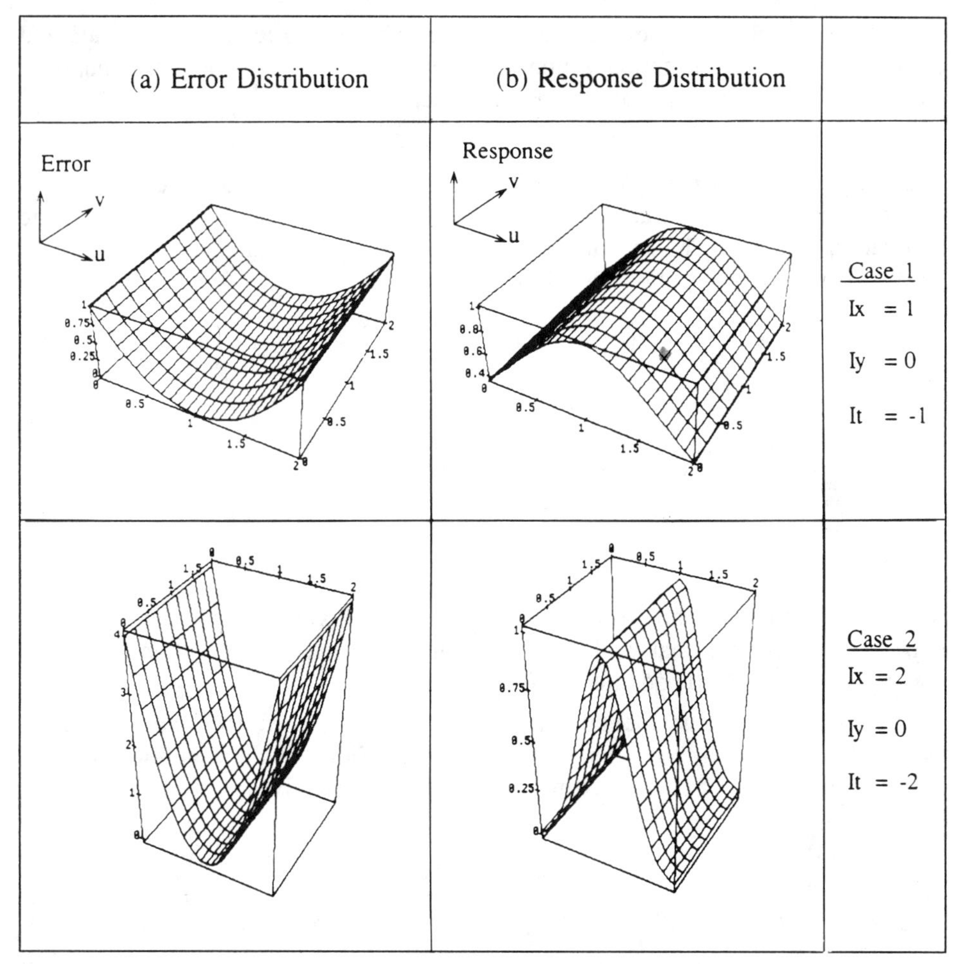

Figure 8.3. Error distribution (a) and response distribution (b) in gradient-based estimation of velocity in the 2D case.

$$\mathcal{R}_g(u, v) = e^{-k\left[(I_x u + I_y v + I_t)^2\right]} \tag{8.4}$$

Figure 8.3b shows the general nature of response distribution in velocity space. The shape of response distribution is that of a ridge—the higher the spatial gradient, the steeper the ridge. I have shown error distributions and response distributions for two different values of spatial gradient. As expected, the ridge in the response distribution becomes steeper as the magnitude of the spatial gradient increases.

We have to be able to compute the values of spatial and temporal gradients and establish the value of the normalization parameter k to compute response distribution. I discuss these implementation issues in a later section.

Interpreting response distribution. In Chapter 5, I discussed the rationale for interpreting response distribution as a frequency distribution in velocity space. The same rationale applies here. The estimate of velocity, $U_{cg} = (u_{cg}, v_{cg})$, computed by a weighted least-squares approach, is given by [144]

$$u_{cg} = \frac{\sum_u \sum_v \mathcal{R}_g(u, v)u}{\sum_u \sum_v \mathcal{R}_g(u, v)}$$

$$v_{cg} = \frac{\sum_u \sum_v \mathcal{R}_g(u, v)v}{\sum_u \sum_v \mathcal{R}_g(u, v)} \tag{8.5}$$

where the summation is carried out over $-N \leq u, v \leq +N$. This estimate can also be thought of as the initial estimate that would serve as input to the velocity-propagation procedure, making use of neighborhood information.

Furthermore, we can assume additivity, zero mean, and independence about the errors, as in the case of the correlation-based approach. With these assumptions, we can associate the following covariance matrix with the estimate given above [144]:

$$S_{cg} = \begin{pmatrix} \frac{\sum_u \sum_v \mathcal{R}_g(u, v)(u - u_{cg})^2}{\sum_u \sum_v \mathcal{R}_g(u, v)} & \frac{\sum_u \sum_v \mathcal{R}_g(u, v)(u - u_{cg})(v - v_{cg})}{\sum_u \sum_v \mathcal{R}_g(u, v)} \\ \frac{\sum_u \sum_v \mathcal{R}_g(u, v)(u - u_{cg})(v - v_{cg})}{\sum_u \sum_v \mathcal{R}_g(u, v)} & \frac{\sum_u \sum_v \mathcal{R}_g(u, v)(v - v_{cg})^2}{\sum_u \sum_v \mathcal{R}_g(u, v)} \end{pmatrix} \tag{8.6}$$

Again, the summation is carried out over $-N \leq u, v \leq +N$.

Alternatively, we could represent the response distribution with a pair of mutually orthogonal conservation constraints, each with a confidence measure. These constraints, being the principal axes of response distribution $\mathcal{R}_g(u, v)$, can be computed simply as the eigenvectors of the covariance matrix defined above and can be written in the following form [105, 135]:

$$\mathcal{L}_{1g} : a_{1g}u + b_{1g}v + c_{1g} = 0$$

$$\mathcal{L}_{2g} : a_{2g}u + b_{2g}v + c_{2g} = 0 \tag{8.7}$$

The two constraints intersect at $U_{cg} = (u_{cg}, v_{cg})$. We can compute the confidence measures C_{1g} and C_{2g} associated with these constraints easily as reciprocals of the normalized moments of inertia of the response distribution about these two lines. The normalized moments of inertia are equal to the squares of eigenvalues of the covariance matrix. However, only one of the two confidence measures, C_{1g} (associated with the axis along the ridge), is reliable. This is because response distribution does not have a well-defined peak; it only has a ridge. Therefore, it is preferable to associate a very low value with the confidence measure C_{2g}. In the implementation used in this book, C_{2g} is explicitly assigned a value of zero.

Some implementation issues. Some issues that relate to implementation of the procedure described above must be addressed. They are

- computation of the spatial and temporal gradients of (low-pass filtered) intensity,
- selection of the normalization parameter k used in converting error distribution into response distribution, and
- quantification of the response distribution.

The following discussion addresses these issues.

Computation of the spatial and temporal gradients. I use $3 \times 3 \times 3$ neighborhoods to compute the partial derivatives of intensity according to the technique proposed by Beaudet [154]. He described a set of masks to compute the partial derivatives at a point based on an $n \times n$ window around it. His approach was based on approximating the image function at a point by a Taylor series and computing the Taylor coefficients to minimize the error between the actual value of the function and its Taylor approximation. The coefficients thus obtained are the various partial derivatives of the image function. Beaudet was able to express these coefficients as a weighted sum of the function values in the $n \times n$ window, thus obtaining a mask to compute the partial derivatives. These masks have been reliably used in various past works on optic-flow computation [68, 80].

Selection of the normalization parameter. The criteria for selecting the value of normalization parameter k have been discussed for the correlation-based approach. The same criteria apply here. Therefore, to establish the value of k, I use the procedure outlined in the section in Chapter 5 on conservation information implementation issues.

Quantification of the response distribution. As shown in the earlier section describing the computational framework, response distribution is available in an analytic form (as a function of u and v). This gives a greater flexibility in computation of the estimate U_{cg} and the covariance matrix S_{cg} as compared with the correlation-based approach. There are three possibilities: discrete computation, analytic solution, and table lookup. I describe each below, assuming that any component of the true velocity is less than two pixels per frame. Hence, using a window given by $-5 \leq u, v \leq +5$ in velocity space is sufficient.

In discrete computation, response distribution can be evaluated at some regularly spaced discrete points in the window $-5 \leq u, v \leq +5$, say, 25 points at all the integer values of u and v. Computation of U_{cg} and S_{cg} from these 25 discrete values can be done in exactly the same way as in the correlation-based approach. All these computations have to be done *on line*.

In analytic computation, the elements of U_{cg} and S_{cg} can be expressed *off line* as analytic functions of I_x, I_y, and I_t (by replacing the summation in Equation 8.6 with double integration). The only computation that has to be done on line is substitution of the actual values of I_x, I_y, and I_t at the point under consideration to obtain the actual values of the elements. This implementation is more efficient than discrete computation.

On the basis of the analytic solution obtained above, lookup tables can be generated off line for regularly spaced values of I_x, I_y, and I_t. For example, assuming that each of these derivatives could lie between −128 and +127 and using only the even values, a table of 128 × 128 × 128 entries will be required. The only computation that has to be done on line is the lookup of the appropriate entry in the table. This is the most efficient implementation. However, I use discrete computation for simplicity.

Figure 8.4 shows the estimates for the running example. Three successive 128 × 128 images of the image sequence are used. One of the original images and the corresponding low-pass filtered version are shown in Figures 8.4a and 8.4b, respectively. Figure 8.4c shows the image depicting the confidence measure C_{1g} at each point. Figure 8.4d shows the velocity estimates (u_{cg}, v_{cg}) over the image.

Relationship with major current approaches. My approach is inspired primarily by the work of Horn and Schunck [34]. There are, however, some significant differences between my approach and theirs:

- Whereas Horn and Schunck compute the (normal component) of velocity by taking the quotient of the temporal and spatial gradients, I compute it as the center of mass of the response distribution in velocity space.
- An explicit notion of confidence measures is missing in Horn and Schunck's model. In my model, the locally available velocity information is recovered in the form of conservation constraints accompanied by their confidence measures.

Conservation information: Spatiotemporal energy-based approach

The organization of this section is similar to that of the previous section. After a brief qualitative description of the overall procedure, I present the quantitative details and discuss various implementation issues. I also discuss the relationship of this framework with some existing ones.

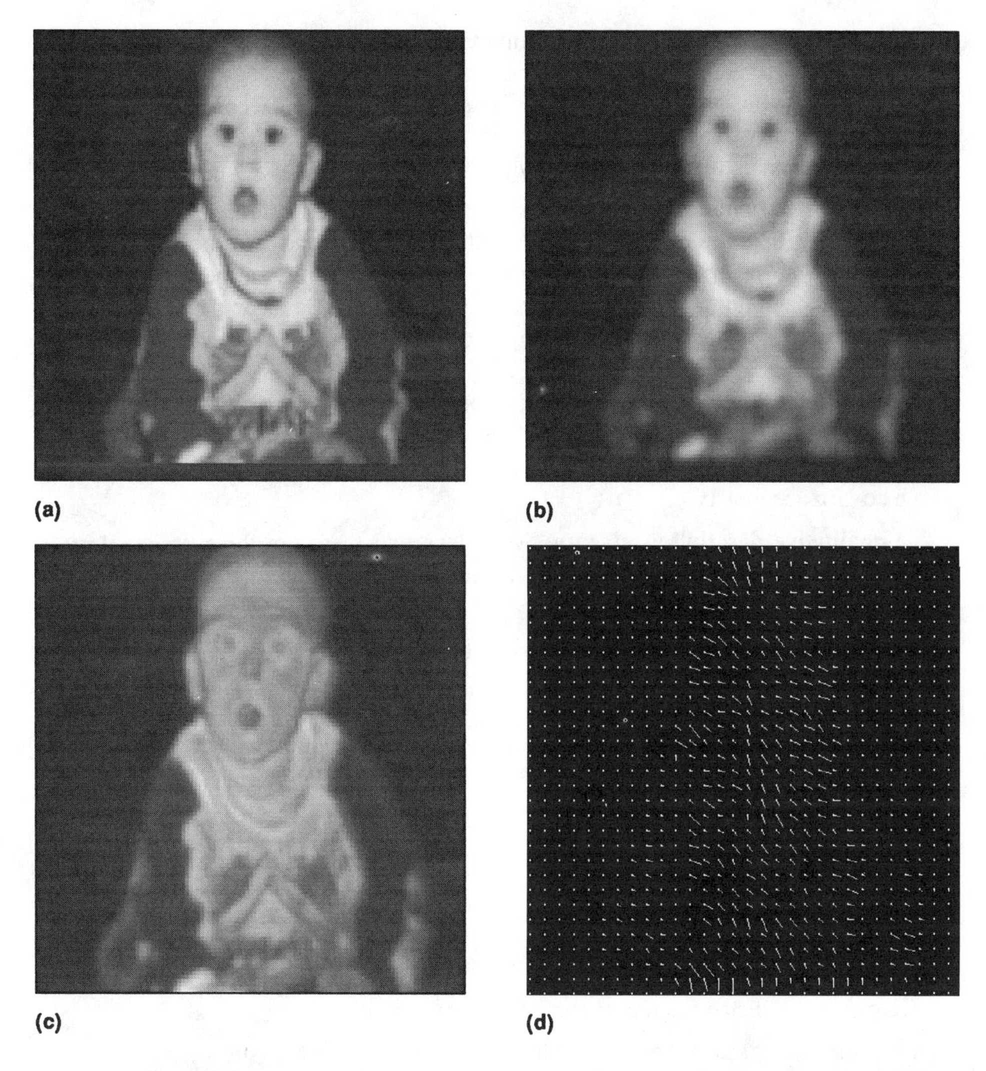

Figure 8.4. An illustration of gradient-based velocity estimates and confidence measures for the running example: (a) one of the images, (b) low-pass filtered version of the image, (c) confidence measure C_{1g}, (d) estimates of velocity (u_{cg}, v_{cg}).

A qualitative description. As discussed in Chapter 3, the problem of local recovery of velocity information can be posed as that of identifying a plane in spatiotemporal frequency space that has the maximum energy associated with it. Typically, this is done by performing a discrete sampling of the motion-energy distribution (of a small spatio-temporal neighborhood around the pixel under consideration) over spatiotemporal frequency space by using a set of spatiotemporally tuned filters [88]. (As explained in Chapter 3, computation of motion energy is done on the imagery corresponding to the conserved quantity. Heeger, for example, assumed conservation of intensity and hence used the original intensity imagery to compute motion energy [88].) After the discrete sampling is performed, the parameters of the plane are computed to give the best fit to the sampled distribution, that is, to minimize some error norm. According to Equation 3.6, these parameters are u and v, the two components of velocity. This scheme, therefore, gives an estimate of velocity without any measure of confidence.

I extend this scheme to incorporate an explicit notion of confidence measures as follows. Rather than search for the velocity that minimizes the error norm, I compute the magnitude of the error norm at various points in the vicinity of the expected value of velocity, thus obtaining an error distribution in uv space. If the expected velocity is not known, the error norm could be computed for several closely spaces points (u, v) such that $-N \leq u, v \leq +N$, where N is the maximum possible magnitude of any component of velocity. I then convert this error distribution into a response distribution in uv space. It follows from the description of the correlation-based and gradient-based approaches that the response distribution can be interpreted to obtain an estimate of velocity along with a covariance matrix.

The qualitative description given above shows that there are three essential steps underlying the recovery of conservation information using the spatiotemporal energy-based approach:

1. Select the conserved quantity and derive it from intensity imagery.
2. Compute error distribution and response distribution in velocity space.
3. Interpret response distribution.

Computation of error distribution and response distribution is sufficiently complex to justify further subdivision into the following four steps:

1. Select a mechanism to sample the motion-energy distribution over spatiotemporal frequency space.
2. Sample the motion-energy distribution.
3. Transform the motion-energy distribution into error distribution over velocity space.
4. Convert error distribution into response distribution.

I show the computational details of these steps in the following subsection.

The computational framework. The overall computational procedure is based on the model proposed by Heeger [88], and good understanding of his model is assumed in the following discussion.

Selecting the conserved quantity and computing it from intensity imagery. I use band-pass filtered intensity as the conserved quantity. The reason for this choice is the following. Band-pass filtering limits the range of spatial and temporal frequencies that can appear in the output imagery. This assists in selecting the optimum values of the parameters (such as frequency tuning and spread) of the filters used to sample the motion-energy distribution. In a full-fledged scheme, we could construct a band-pass pyramid, each level of which has a different range of spatial frequencies. Further, we could use a distinct set of spatiotemporally-oriented filters to sample the motion-energy distribution at each level, thus covering the entire range of frequencies. In the discussion that follows, I use only one level of the pyramid (at highest resolution) to illustrate the procedure. Computation of band-pass imagery has been discussed in the case of the correlation-based approach. I use the same implementation here. For brevity, I use the term "image" to refer to "band-pass filtered image."

Computing error distribution and response distribution in velocity space. First, we select a mechanism for sampling the motion-energy distribution. The discussion in Chapter 3 showed that the "hardware" needed to estimate optic flow using the spatiotemporal energy-based approach is a set of spatiotemporally-oriented filters. The frequency response of these filters is high in the vicinity of a preferred spatiotemporal frequency and decays away from it. In other words, these filters are tuned to specific spatiotemporal frequencies and can be used to sample the motion energy of a time-varying stimulus at these frequencies.

In the past research, several mechanisms have been proposed to construct spatiotemporally-oriented filters. Fleet and Jepson [87] used a linear combination of several difference-of-Gaussian (DOG) filters spatially and temporally shifted with respect to each other. They showed that the combination is tuned to a band of spatiotemporal frequencies. The location and the extent of the band is controlled by the spread and the offset of the individual DOG filters. Gabor filters [155] constitute another class of spatiotemporally-oriented filters. Heeger [88] showed that they can be used to sample motion energy and described a technique to recover optic flow using them. Mallat [156] showed that spatial quadrature-mirror filters can be used to construct mechanisms that exhibit spatial orientation selectivity. It appears feasible that a temporal extension of his framework can serve as a motion-energy sampler. In an earlier work [157], I showed a class of spatially oriented filters implemented as a sequence of Laplacian and Gaussian operations. I also alluded to its temporal extension, which might exhibit spatiotemporal orientation selectivity.

Although the mechanisms summarized above exhibit spatiotemporal orientation selectivity, none except Gabor filters has actually been used to recover optic flow. More research needs to be done to evaluate their relative merits. In the model discussed below, I use Gabor filters as the motion-energy sampling mechanism.

The next step is to sample the motion-energy distribution over spatiotemporal frequency space. A typical space-time Gabor filter is represented as the product of a space-time Gaussian and a sine (or cosine) wave and is given by

$$g(x, y, t) = \frac{1}{\sqrt{2}\pi^{1.5}\sigma_x\sigma_y\sigma_t} \exp\left\{ -\left[\frac{x^2}{2\sigma_x^2} + \frac{y^2}{2\sigma_y^2} + \frac{t^2}{2\sigma_t^2}\right]\right\} \sin(2\pi\omega_{x_0} + 2\pi\omega_{y_0} + 2\pi\omega_{t_0}) \quad \textbf{(8.8)}$$

where $\omega_{x_0}, \omega_{y_0}, \omega_{t_0}$ is the *center frequency*, that is, the spatial and temporal frequency for which the filter gives its maximum response, and $(\sigma_x, \sigma_y, \sigma_t)$ is the spread of the spatiotemporal Gaussian. The frequency response of this filter is given by

$$G(\omega_x, \omega_y, \omega_t) = \frac{1}{4}\exp\{-4\pi^2[\sigma_x^2(\omega_x - \omega_{x_0})^2 + \sigma_y^2(\omega_y - \omega_{y_0})^2 + \sigma_t^2(\omega_t - \omega_{t_0})^2]\}$$
$$+\frac{1}{4}\exp\{-4\pi^2[\sigma_x^2(\omega_x + \omega_{x_0})^2 + \sigma_y^2(\omega_y + \omega_{y_0})^2 + \sigma_t^2(\omega_t + \omega_{t_0})^2]\} \quad \textbf{(8.9)}$$

This implies that a filter with center frequency $(\omega_{x_0}, \omega_{y_0}, \omega_{t_0})$ will give an output equal to $G(\omega_x, \omega_y, \omega_t)$ when stimulated by (i.e., convolved with) a signal with a spatiotemporal frequency $(\omega_x, \omega_y, \omega_t)$. The sum of the squared outputs of the sine-phase and cosine-phase Gabor filters (both having the same center frequency) gives the motion energy. This sum will be large if the motion energy of the input stimulus is concentrated around $(\omega_{x_0}, \omega_{y_0}, \omega_{t_0})$, and it will be small if the motion energy is concentrated at a frequency away from $(\omega_{x_0}, \omega_{y_0}, \omega_{t_0})$ [88].

Thus, a Gabor filter can be used to sample the energy content of a small spatiotemporal neighborhood at a given frequency: the center frequency of the filter. This can be done by convolving the spatiotemporal neighborhood of interest with space-time sine-phase and cosine-phase Gabor filters, squaring the outputs and computing their sum. In other words, we can use *a set of* Gabor filters, tuned to various frequencies in the spatiotemporal frequency space, to sample the entire distribution of the motion energy of a time-varying stimulus over the spatiotemporal frequency space.

Exactly how many filters should there be in this set? How should the center frequencies and the spreads of these filters be selected? More research needs to be done to establish definitive answers to these questions. I use Heeger's configuration: 12 filters with identical spread, tuned to center frequencies that lie along a cylinder in spatiotemporal frequency space. The axis of the cylinder is the ω_t axis. Figure 8.5 shows the power spectra of the 12 files. I denote the energy sampled by the ith Gabor filter by E_{s_i}.

It is also possible to *predict* the energy output of a Gabor filter in response to a stimulus whose spatial frequency and velocity are known (using Equation 8.9 and replacing ω_t with $\omega_x u + \omega_y v$ according to Equation 3.6). Heeger [88] showed that the predicted Gabor energy of a small spatiotemporal neighborhood in a random-dot texture translating with a velocity (u, v), sampled by the ith Gabor filter, is given by

$$E_{p_i}(u, v) = \exp\left[-4\pi^2\sigma_x^2\sigma_y^2\sigma_t^2 \frac{(u\omega_{x_i} + v\omega_{y_i} + \omega_{t_i})^2}{(v\sigma_x\sigma_t)^2 + (u\sigma_y\sigma_t)^2 + (\sigma_x\sigma_y)^2}\right] \quad \textbf{(8.10)}$$

where $(\omega_{x_i}, \omega_{y_i}, \omega_{t_i})$ is the center frequency of the filter and $(\sigma_x, \sigma_y, \sigma_t)$ is the spread.

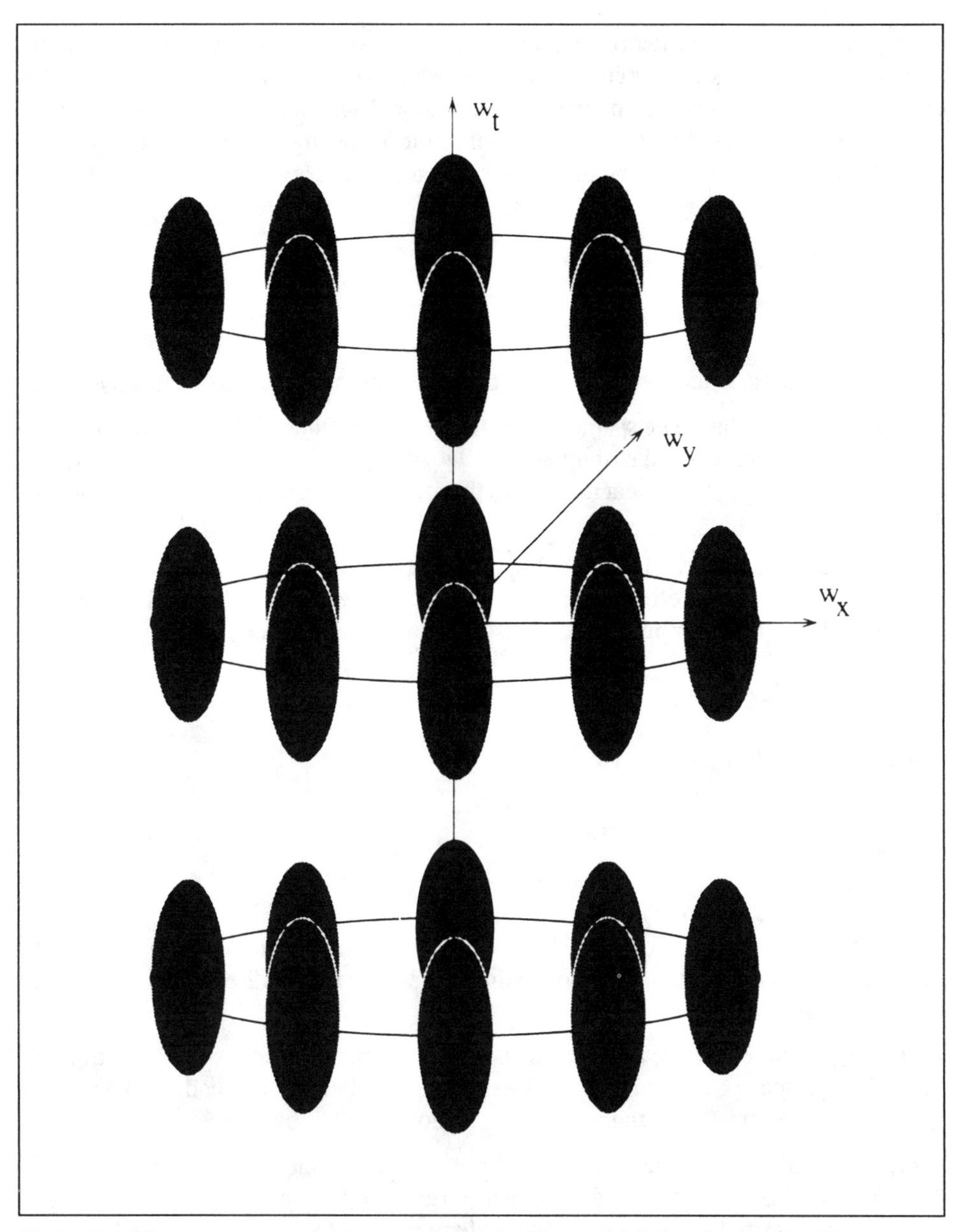

Figure 8.5. The power spectra of 12 spatiotemporally tuned Gabor filters. Each filter has a power spectrum of the form of a pair of ellipsoids (3D Gaussians) centered around $(\omega_{x_0}, \omega_{y_0}, \omega_{t_0})$ and $(-\omega_{x_0}, -\omega_{y_0}, -\omega_{t_0})$.

Next, we compute the error distribution in velocity space. At this point, we have two measures of the motion-energy distribution of a small spatiotemporal neighborhood in an image sequence: the actual energy sampled by each of the 12 filters, and the predicted energy for each of the 12 filters expressed in terms of the correct velocity (u, v), assuming that the time-varying stimulus can be locally approximated by a random-dot texture. The correct velocity (u, v) is the one that minimizes the discrepancy between the sampled energy and the predicted energy. Heeger defined the following error function to quantify the discrepancy between the sampled and the predicted energy:

$$\sum_{i=1}^{12} \left[E_{s_i} - \overline{E_{s_i}} \frac{E_{p_i}(u, v)}{\overline{E_{p_i}(u, v)}} \right]^2 \tag{8.11}$$

where $\overline{E_{s_i}}$ and $\overline{E_{p_i}(u, v)}$ are the sums of sampled and predicted energies, respectively, for the filters that have the same spatial orientation as the ith filter [88]. This error function essentially gives an error distribution over uv space. Heeger performed an iterative search for the velocity (u, v) that minimizes the error. It is at this point that my approach differs from Heeger's.

To convert error distribution into response distribution, rather than search for the global minimum of the error function, I define a *response distribution* by using an exponential function (for reasons discussed in the case of the correlation-based approach):

$$\mathcal{R}_e(u, v) = e^{-k \sum_{i=1}^{12} \left[E_{s_i} - \overline{E_{s_i}} \frac{E_{p_i}(u, v)}{\overline{E_{p_i}(u, v)}} \right]^2} \qquad -N \leq u, v \leq +N \tag{8.12}$$

Then I plot the distribution over a $(2N + 1) \times (2N + 1)$ window around $(0, 0)$ in velocity space, under the assumption that the maximum velocity in any direction is less than N pixels per frame.

From the empirical observations made below, it is apparent that response distribution can be interpreted in exactly the same way as in the correlation-based and gradient-based approaches—to obtain an estimate of velocity along with a covariance matrix.

- If the underlying spatiotemporal neighborhood has sufficient isotropic texture (as in a random-dot field), the predicted energy (given by Equation 8.10) is exact. The response distribution in uv space has a sharp peak at the true velocity and falls off quickly. This is depicted in Figure 8.6a.
- If the contrast in the spatiotemporal neighborhood used above is decreased, the peak in the response distribution becomes fuzzy. That is, the falloff is more gradual, as shown in Figure 8.6b.
- If the contrast in the spatiotemporal neighborhood is made anisotropic (for example, by having two different contrasts in two orthogonal spatial directions), the peak in

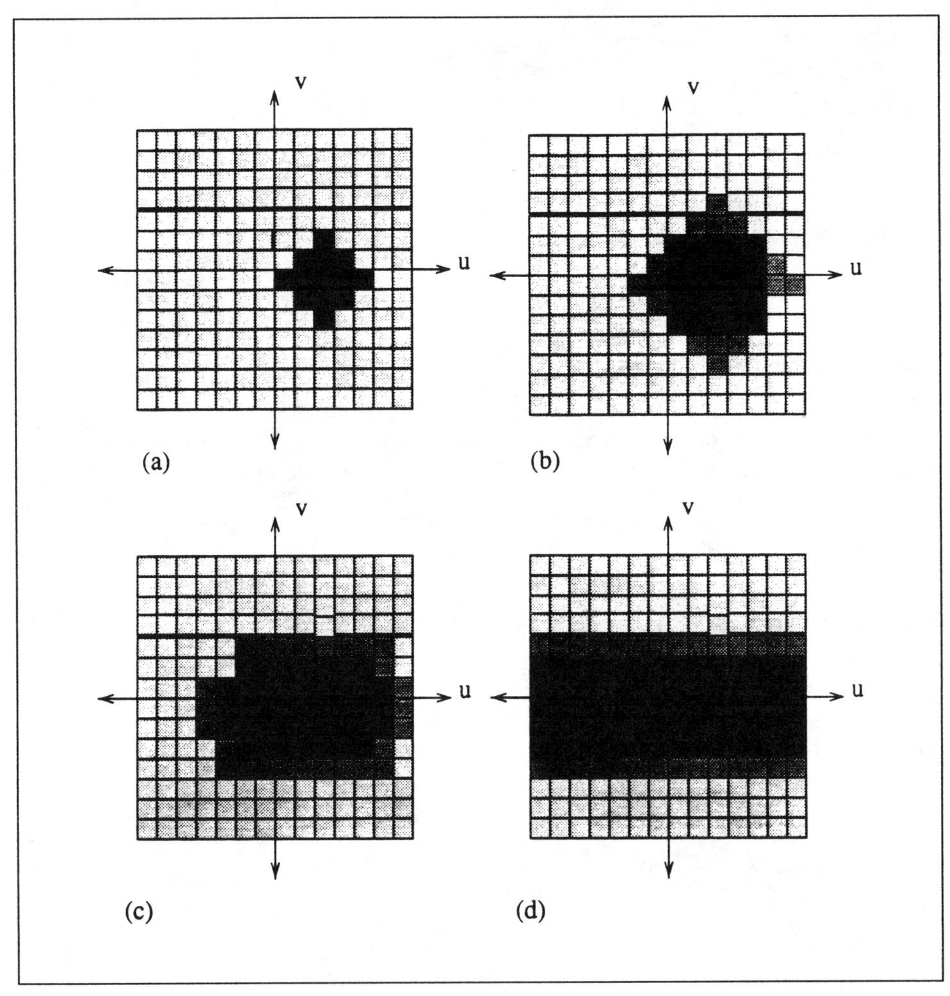

Figure 8.6. Behavior of the response distribution in velocity space.

the response distribution tends to become elongated. That is, the falloff is sharper in one direction than in the orthogonal direction, as depicted in Figure 8.6c.

- If the contrast in one spatial direction completely vanishes, the peak in the response distribution takes the form of a ridge. This is shown in Figure 8.6d.

The quantitative procedure for interpreting response distribution is described below.

Interpreting response distribution. Following the arguments given for the correlation-based and gradient-based approaches, the weighted least-squares estimate of velocity, $U_{ce} = (u_{ce}, v_{ce})$, is given by

$$u_{ce} = \frac{\sum_u \sum_v \mathcal{R}_e(u, v)u}{\sum_u \sum_v \mathcal{R}_e(u, v)}$$

$$v_{ce} = \frac{\sum_u \sum_v \mathcal{R}_e(u, v)v}{\sum_u \sum_v \mathcal{R}_e(u, v)} \tag{8.13}$$

where the summation is carried out over $-N \le u, v \le +N$.

Furthermore, we can assume additivity, zero mean, and independence about measurement errors, as in the case of the correlation-based and gradient-based approaches. With these assumptions, we can associate the following covariance matrix with the estimate given above [144]:

$$S_{ce} = \begin{pmatrix} \dfrac{\sum_u \sum_v \mathcal{R}_e(u, v)(u - u_{ce})^2}{\sum_u \sum_v \mathcal{R}_e(u, v)} & \dfrac{\sum_u \sum_v \mathcal{R}_e(u, v)(u - u_{ce})(v - v_{ce})}{\sum_u \sum_v \mathcal{R}_e(u, v)} \\ \dfrac{\sum_u \sum_v \mathcal{R}_e(u, v)(u - u_{ce})(v - v_{ce})}{\sum_u \sum_v \mathcal{R}_e(u, v)} & \dfrac{\sum_u \sum_v \mathcal{R}_e(u, v)(v - v_{ce})^2}{\sum_u \sum_v \mathcal{R}_e(u, v)} \end{pmatrix} \tag{8.14}$$

Again, the summation is carried out over $-N \le u, v \le +N$.

Alternatively, we could represent the response distribution with a pair of mutually orthogonal conservation constraints, each with a confidence measure. These constraints, being the principal axes of response distribution $\mathcal{R}_e(u, v)$, can be computed simply as the eigenvectors of the covariance matrix defined above and can be written in the following form [105, 135]:

$$\mathcal{L}_{1e} : a_{1e}u + b_{1e}v + c_{1e} = 0$$

$$\mathcal{L}_{2e} : a_{2e}u + b_{2e}v + c_{2e} = 0 \tag{8.15}$$

The two constraints intersect at $U_{ce} = (u_{ce}, v_{ce})$. We can easily compute their associated confidences C_{1e} and C_{2e} as reciprocals of the normalized moments of inertia of the response distribution about these two lines. The normalized moments of inertia are equal to the squares of the eigenvalues of the covariance matrix.

Some implementation issues. There are two important issues related to implementation of the procedure described above: implementation of Gabor filters and selection of the window size in uv space to compute response distribution.

Implementation of Gabor filters. As I mentioned earlier, further research needs to be done to establish definitive answers to the questions concerning selection and implementation of Gabor filters. For this reason, I use the same filters as those used by Heeger and implement them using his space-time separable method. The spatial frequency tuning of each filter is at 0.25 cycles per pixel and the temporal frequency tuning is either at 0 cycles per frame or at 0.25 cycles per frame. The standard deviation of all spatial Gaussians is $\sigma_x = \sigma_y = 4$ and that of temporal Gaussians is $\sigma_t = 1$. Motion energy is computed easily as the sum of squared outputs of the sine-phase and cosine-phase filters.

Selection of window size in velocity space. I use a 5×5 window (25 discrete points) centered at (0, 0) in uv space to plot the response distribution. As with the correlation-based and gradient-based approaches, this choice is based on the assumption that the maximum possible magnitude of any component of velocity is less than two pixels. Also, the normalization parameter k is established in exactly the same way as in the correlation-based approach.

Figure 8.7 shows the results for the running example. Seven successive 128×128 images of the image sequence are used. Figure 8.7a shows the band-pass filtered version of one of the images. Figures 8.7b and 8.7c show respectively images depicting the confidences C_{1e} and C_{2e} at each point. Figure 8.7d shows the velocity estimates (u_{ce}, v_{ce}) over the image.

Relationship with major current approaches. My approach is primarily motivated by the work of Heeger [88]. I use the same hardware as Heeger—that is, a set of spatiotemporally tuned Gabor filters—to sample the distribution of motion energy of a small spatiotemporal neighborhood over spatiotemporal frequency space. Also, I transform the energy distribution in spatiotemporal frequency space into the response distribution in velocity space in a manner similar to Heeger's. However, I use response distribution differently. Whereas Heeger searches for the global maximum of response distribution, I compute its center of mass and use it as the velocity estimate. Furthermore, I compute the covariance matrix associated with this estimate on the basis of the falloff characteristics of the response distribution.

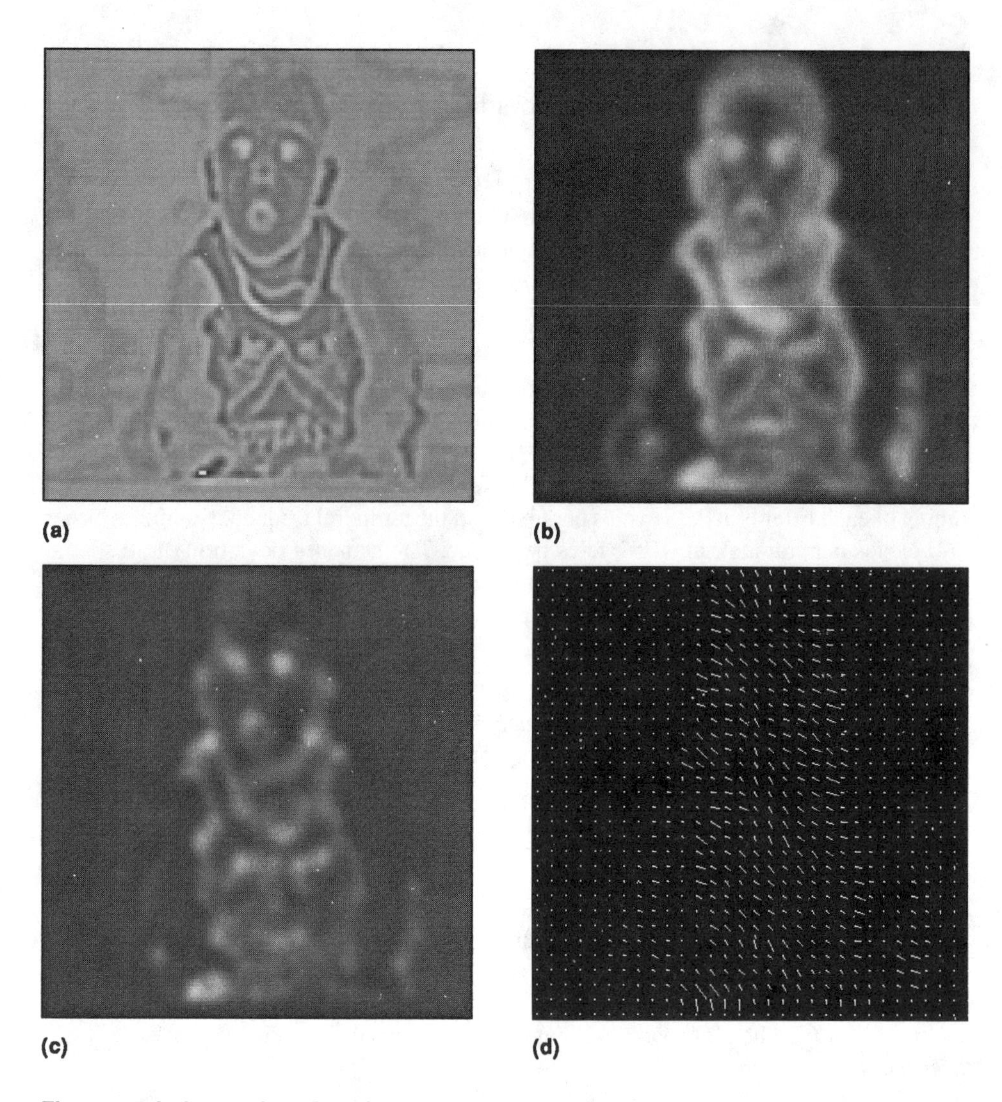

Figure 8.7. An illustration of spatiotemporal energy-based velocity estimates and confidence measures for the running example: (a) band-pass filtered version of one of the images, (b, c) confidence measures C_{1e} and C_{2e}, respectively, (d) estimates of velocity (u_{ce}, v_{ce}).

Neighborhood information

I show in this section that the velocity-propagation technique developed in my framework reduces to conventional smoothing [34, 80] under certain specific operating conditions. I begin this discussion by reminding the reader that the velocity-propagation technique used here originates from the following formulation of neighborhood error:

$$(U - \overline{U})^T S_n^{-1} (U - \overline{U}) \tag{8.16}$$

where $\overline{U}$ denotes the (weighted) average velocity in the neighborhood, and S_n denotes the covariance matrix representing the neighborhood velocity distribution.

I describe the relationship of the neighborhood error shown above to the corresponding "smoothness error" used by Anandan [80] and Horn and Schunck [34]. The expression for smoothness error is given by

$$\alpha^2((u - \overline{u})^2 + (v - \overline{v})^2) \tag{8.17}$$

which can be rewritten in matrix notation as

$$(U - \overline{U})^T (\alpha^2 I)(U - \overline{U}) \tag{8.18}$$

where I is the 2×2 identity matrix and α is the smoothness factor. It is now apparent that Horn and Schunck's smoothness error can be derived from the neighborhood error used in this framework, simply by replacing S_n^{-1} with $\alpha^2 I$.

A quick glance at Figure 4.4a reveals that the neighborhood velocity distribution in homogeneous regions has the form of a single cluster whose two principal axes have almost identical "spreads." Furthermore, without loss of generality, we can assume that the principal axes are aligned with the axes of the coordinate system. Therefore, the scatter matrix S_n is an identity matrix (up to a scale factor). Hence, S_n^{-1} is indeed of the form $\alpha^2 I$. Thus, in view of my framework, the formulation of smoothness error is valid in homogeneous regions. Various propagation procedures using conventional smoothing work well exactly in such regions [34, 80, 83].

Conclusion

In this chapter, I have shown that the framework discussed in Chapter 5 applies identically to the correlation-based, gradient-based, and spatiotemporal energy-based approaches for the first step in optic-flow estimation, that is, recovering conservation information. Alternatively, the three basic approaches can be unified in light of the estimation-theoretic framework. In the next chapter, I show that this unification leads naturally to a framework to integrate the three basic approaches. I have also shown that the technique for velocity propagation—the second step in optic-flow estimation—used in this framework reduces to conventional smoothing under certain specific operating conditions. In other words, the estimation-theoretic framework also unifies some existing approaches for velocity propagation.

Chapter 9

Integration

In the context of this book, integration refers to a computational framework where the three basic approaches for recovering local velocity information work in cooperation to give a robust initial estimate. This initial estimate is then passed through the velocity-propagation procedure to yield the final flow field. In this chapter, I discuss the issue of integration. As I mentioned in the overview in Chapter 1, the objective of this chapter is simply to show the potential advantage of integrating the three existing approaches and to show that the framework developed in Chapters 4 and 5 appears to be a suitable platform to accomplish integration. Further research needs to be done to understand the various implementation issues and to determine the overall advantage of the integrated scheme.

First, I discuss the motivation for integration and show some mathematical preliminaries. Then, I present algorithms based on the integrated framework.

Motivation

The motivation for integration can be explained as follows. As discussed in Chapters 4, 5, and 8, the recovery of locally available velocity information is never error free. Furthermore, because of inherent differences in the measurements used by the three approaches, the error characteristics of the estimates obtained from them are not identical. For illustration, Table 9.1 shows the results obtained at point P in the running example (see Figure 5.2) by using each of the three approaches. It is apparent that the estimates and the covariance matrices (i.e., the error characteristics) obtained from the three approaches differ. This scenario is representative of the classic multisensor problem [158]. Algorithms based on the three approaches can be thought of as three different sensors measuring a given quantity—optic flow—with different error characteristics. The measurements from different sensors can be fused MSE-optimally to produce an initial estimate of optic flow that serves as an input to the velocity-propagation procedure. The notion of MSE-optimality of the estimate can be understood from the following example, which is a variant of a multisensor fusion problem discussed by Gelb [158].

Example. Consider a system composed of three sensors, making measurements m_1, m_2, and m_3 of an unknown quantity x, in the presence of random, independent, unbiased

Table 9.1. An illustration of the error characteristics of optic-flow estimates obtained from three approaches.				
		Correlation-Based Approach	Gradient-Based Approach	Energy-Based Approach
Velocity Estimate	u	0.356	0.296	0.432
	v	–0.418	–0.212	–0.392
Covariance Matrix	S_{11}	1.461	0.072	0.939
	S_{12}	0.494	0.023	0.621
	S_{22}	1.028	1.691	0.816

measurement errors e_1, e_2, and e_3, respectively, with variances given by σ_1^2, σ_2^2, and σ_3^2, respectively. The objective is a scheme to fuse the measurements to produce an MSE-optimal estimate of x, where MSE-optimality means that

- the mean value of error in the resultant estimate is equal to zero, and
- the mean-squared error in the resultant estimate is minimum.

Let the estimate of x (denoted by $\hat{x}$) be given by a linear combination of the individual measurements as follows:

$$\hat{x} = k_1 m_1 + k_2 m_2 + k_3 m_3 \tag{9.1}$$

The coefficients k_1, k_2, and k_3 must be determined to satisfy the following two criteria of MSE-optimality:

$$E[\tilde{x}] = 0 \tag{9.2}$$

$$E[\tilde{x}^2] = \text{Minimum} \tag{9.3}$$

where E denotes the mean and $\tilde{x}$ refers to the estimation error given by $\hat{x} - x$.

From the first criterion (Equation 9.2), we can write

$$E[\tilde{x}] = E[k_1 m_1 + k_2 m_2 + k_3 m_3 - x] = 0 \tag{9.4}$$

Using the property that the mean value of a linear combination of several random variables is equal to the same linear combination of the mean values of the random variables, we can write [144]

$$E[k_1(x+e_1)+k_2(x+e_2)+k_3(x+e_3)-x]=0 \quad \textbf{(9.5)}$$

or

$$(k_1+k_2+k_3-1)E[x]+k_1E[e_1]+k_2E[e_2]+k_3E[e_3]=0 \quad \textbf{(9.6)}$$

Since e_1, e_2, and e_3 are unbiased, the mean value of each is zero. Also, since x is nonrandom, its mean value is x. Thus, Equation 9.6 gives

$$k_3=1-k_1-k_2 \quad \textbf{(9.7)}$$

From the second criterion (Equation 9.3), we can write

$$E[\tilde{x}^2]=E[(k_1(x+e_1)+k_2(x+e_2)+k_3(x+e_3)-x)^2]=\text{Minimum} \quad \textbf{(9.8)}$$

Again, using the linear combination principle mentioned above and substituting the value of k_3 from Equation 9.7, we can rewrite this as

$$k_1^2E[e_1^2]+k_2^2E[e_2^2]+(1-k_1-k_2)^2E[e_3^2]=\text{Minimum} \quad \textbf{(9.9)}$$

which implies that

$$k_1^2\sigma_1^2+k_2^2\sigma_2^2+(1-k_1-k_2)^2\sigma_3^2=\text{Minimum} \quad \textbf{(9.10)}$$

For the mean-squared error to be minimum, its partial derivatives with respect to k_1 and k_2 must be zero. That is,

$$2k_1\sigma_1^2-2(1-k_1-k_2)\sigma_3^2=0 \quad \textbf{(9.11)}$$

$$2k_2\sigma_2^2-2(1-k_1-k_2)\sigma_3^2=0 \quad \textbf{(9.12)}$$

Solving Equations 9.7, 9.11, and 9.12 simultaneously, we obtain the following expressions for k_1, k_2, and k_3:

$$k_1 = \frac{\sigma_2^2 \sigma_3^2}{\sigma_1^2 \sigma_2^2 + \sigma_2^2 \sigma_3^2 + \sigma_3^2 \sigma_1^2}$$

$$k_2 = \frac{\sigma_3^2 \sigma_1^2}{\sigma_1^2 \sigma_2^2 + \sigma_2^2 \sigma_3^2 + \sigma_3^2 \sigma_1^2} \quad \textbf{(9.13)}$$

$$k_3 = \frac{\sigma_1^2 \sigma_2^2}{\sigma_1^2 \sigma_2^2 + \sigma_2^2 \sigma_3^2 + \sigma_3^2 \sigma_1^2}$$

Thus, the estimate of x and the mean value of its squared error (mean-squared estimation error) can be given by

$$\hat{x} = \frac{1}{\sigma_1^2 \sigma_2^2 + \sigma_2^2 \sigma_3^2 + \sigma_3^2 \sigma_1^2} [\sigma_2^2 \sigma_3^2 m_1 + \sigma_3^2 \sigma_1^2 m_2 + \sigma_1^2 \sigma_2^2 m_3]$$

$$E[\tilde{x}^2] = \left[\frac{1}{\sigma_1^2} + \frac{1}{\sigma_2^2} + \frac{1}{\sigma_3^2} \right]^{-1} \quad \textbf{(9.14)}$$

The mean-squared estimation error is smaller than any one of the three mean-squared measurement errors.

In this example, the quantity to be estimated, $\hat{x}$, is a scalar. However, the concept of fusing measurements from multiple sensors having different error characteristics is quite general and can be extended to vector quantities such as optic flow. I will show a derivation of Equation 9.14 for vector quantities in the following section.

Mathematical preliminaries

The discussion in the previous section reveals that the problem of fusing velocity information obtained from the three basic approaches fits very well in the domain of statistical estimation. The three approaches can be regarded as three sensors, each giving an initial estimate of the velocity vector (u, v) along with a covariance matrix that characterizes the measurement error. The individual estimates can be fused on the basis of their covariance matrices to give an MSE-optimal estimate. I show a framework for fusion in the following theorem. The principles underlying this theorem are well established in statistical estimation theory [144].

Information fusion theorem

Theorem 9.1. There are n sensors, each giving a measurement X_i of an unknown quantity X with an additive, unbiased, and independent error V_i, where X_i, X, and V_i are $p \times 1$ vectors. Each error V_i is characterized by its

$p \times p$ covariance matrix S_i. The optimal estimate $\hat{X}$ for the unknown quantity X and its associated covariance matrix $\hat{S}$ are given as follows. The MSE-optimality is meant to imply that the error in the fused estimate is unbiased and is minimal in a mean-squared sense:

$$\hat{X} = [S_1^{-1} + S_2^{-1} + \dots\dots + S_n^{-1}]^{-1}[S_1^{-1}X_1 + S_2^{-1}X_2 + \dots\dots + S_n^{-1}X_n]$$
$$\hat{S} = [S_1^{-1} + S_2^{-1} + \dots\dots + S_n^{-1}]^{-1} \quad \textbf{(9.15)}$$

I will prove the above theorem for a special case with only two sensors. The extension to n sensors is straightforward. The measurements are related to the unknown quantity as follows:

$$X_1 = X + V_1$$
$$X_2 = X + V_2 \quad \textbf{(9.16)}$$

Using E to denote the mean and using the definition of covariance, we can write

$$E[V_1 V_1^T] = S_1$$
$$E[V_2 V_2^T] = S_2 \quad \textbf{(9.17)}$$

Let $\hat{X}$ be given by a linear combination of X_1 and X_2. That is,

$$\hat{X} = K_1 X_1 + K_2 X_2 \quad \textbf{(9.18)}$$

where K_1 and K_2 are $p \times p$ matrices. The two criteria of MSE-optimality can be written as

$$E[\hat{X} - X] = 0$$
$$E[(\hat{X} - X)(\hat{X} - X)^T] = \text{Minimum} \quad \textbf{(9.19)}$$

Substituting the value of $\hat{X}$ from Equation 9.18 in Equation 9.19 (first criterion), we get

$$E[K_1(X + V_1) + K_2(X + V_2) - X] = 0 \quad \textbf{(9.20)}$$

Using the linear combination principle discussed in the previous section, we rewrite this as

$$(K_1 + K_2 - I)E[X] + K_1E[V_1] + K_2E[V_2] = 0 \quad \textbf{(9.21)}$$

Since V_1 and V_2 are unbiased, their mean values will be zero. Also, since X is nonrandom, its mean value is simply X. Since X is not zero in general, the coefficient of $E[X]$ in Equation 9.21 is zero. Therefore,

$$K_2 = I - K_1 \quad \textbf{(9.22)}$$

Furthermore, substituting the value of $\hat{X}$ from Equation 9.18 in Equation 9.19 (second criterion), we get

$$E[(K_1(X+V_1)+(I-K_1)(X+V_2)-X)(K_1(X+V_1)+ \\ (I-K_1)(X+V_2)-X)^T] = \text{Minimum} \quad \textbf{(9.23)}$$

This gives

$$E[(K_1V_1+(I-K_1)V_2)(K_1V_1+(I-K_1)V_2)^T] = \text{Minimum} \quad \textbf{(9.24)}$$

Using basic matrix algebra, this can be rewritten as [144]

$$E[(K_1V_1+(I-K_1)V_2)(V_1^TK_1^T+V_2^T(I-K_1)^T)] = \text{Minimum} \quad \textbf{(9.25)}$$

Multiplying the two terms and using the principle of linear combination, this reduces to

$$K_1E[V_1V_1^T]K_1^T+(I-K_1)E[V_2V_1^T]K_1^T+ \\ K_1E[V_1V_2^T](I-K_1)^T+(I-K_1)E[V_2V_2^T](I-K_1)^T = \text{Minimum} \quad \textbf{(9.26)}$$

Finally, using Equation 9.17 and the fact that V_1 and V_2 are independent (that is, $E[V_2V_1^T]$ and $E[V_1V_2^T]$ are both zero), we can rewrite Equation 9.26 as

$$K_1S_1K_1^T+(I-K_1)S_2(I-K_1)^T = \text{Minimum} \quad \textbf{(9.27)}$$

For the left-hand side of Equation 9.27 to be minimum, its matrix derivative with respect to K_1 must be zero. That is,

$$\nabla_{K_1}(K_1S_1K_1^T + (I - K_1)S_2(I - K_1)^T) = 0 \tag{9.28}$$

That is,

$$2S_1K_1 - 2S_2(I - K_1) = 0 \tag{9.29}$$

which gives

$$(S_1 + S_2)K_1 = S_2 \tag{9.30}$$

Using the fact that all the matrices involved are symmetrical, the value of K_1 can be derived from this expression as follows:

$$\begin{aligned} K_1 &= (S_1 + S_2)^{-1}S_2 \\ &= (S_2^{-1}(S_1 + S_2))^{-1} \\ &= (S_2^{-1}S_1 + I)^{-1} \\ &= (S_2^{-1}S_1 + S_1^{-1}S_1)^{-1} \\ &= ((S_2^{-1} + S_1^{-1})S_1)^{-1} \\ &= S_1^{-1}(S_1^{-1} + S_2^{-1})^{-1} \end{aligned} \tag{9.31}$$

The value of K_2 can be easily found as follows:

$$K_2 = (I - K_1) = S_2^{-1}(S_1^{-1} + S_2^{-1})^{-1} \tag{9.32}$$

Thus, the MSE-optimal estimate of X for $n = 2$ is given by

$$\hat{X} = (S_1^{-1} + S_2^{-1})^{-1}(S_1^{-1}X_1 + S_2^{-1}X_2) \tag{9.33}$$

The resultant covariance matrix associated with this estimate can be found by substituting the value of K_1 in Equation 9.27. It is given by

$$\hat{S} = (S_1^{-1} + S_2^{-1})^{-1} \quad \textbf{(9.34)}$$

Equations 9.33 and 9.34 prove the theorem for the special case of $n = 2$. The extension for a general value of n is straightforward and is not shown here.

Information fusion and optic flow. The framework for information fusion described above is directly applicable to the problem of integrating various approaches for optic-flow estimation. Using Equation 9.15, we can fuse the initial estimates of optic flow obtained from multiple approaches on the basis of their covariance matrices. This gives a fused estimate and a covariance matrix associated with it. Let the fused estimate be denoted by $U_{cf} = (u_{cf}, v_{cf})$ and its covariance matrix be denoted by S_{cf}. They can now be used as inputs to the velocity-propagation procedure. In other words, they can replace U_c and S_c, respectively, in Equation 8.1, to recover the optic-flow field.

Alternatively, two mutually orthogonal lines passing through (u_{cf}, v_{cf}) and aligned with the eigenvectors of S_{cf} can be thought of as conservation constraints. The confidence measures associated with these constraints can be found as the reciprocals of the eigenvalues of S_{cf}. Conservation constraints can be combined with neighborhood constraints to recover the optic-flow field. This formulation is apparent in the algorithms discussed in the next section.

Algorithms

I will show two algorithms that are functionally equivalent. They are based, respectively, on the two representations for velocity described earlier:

- an estimate accompanied by a covariance matrix, and
- two mutually orthogonal constraints, each with an associated confidence measure.

I refer to these representations as the *first representation* and the *second representation*, respectively, and the corresponding algorithms as Algorithm 3 and Algorithm 4, respectively.

Algorithm 3. This algorithm follows directly from the information fusion theorem. Figure 9.1 gives a schematic description of the algorithm. Apparent in the figure are the two steps in this algorithm: recovery of locally available velocity information, and propagation of velocity information. The computations involved in these two steps are summarized below. Various implementation issues, such as selection of window sizes, filter parameters, and normalization parameters, have already been discussed in Chapters 5 and 8 and are not repeated here.

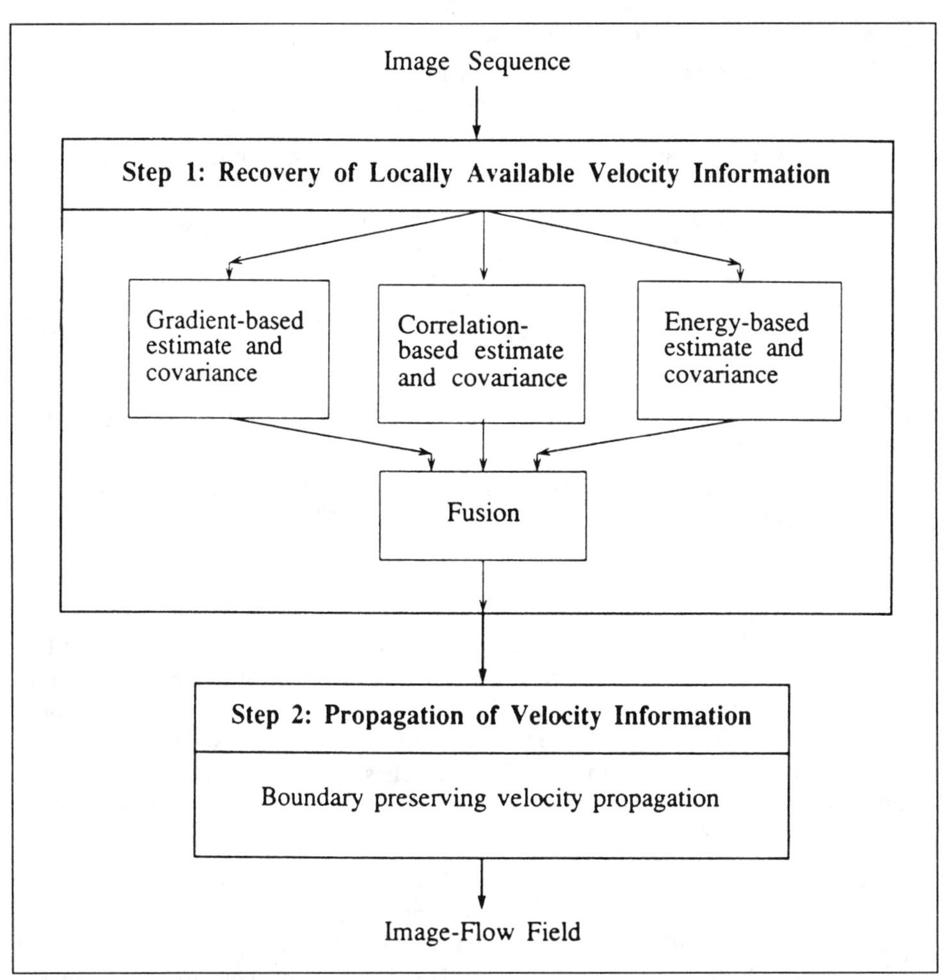

Figure 9.1. An algorithm for extracting optic flow by integrating the correlation-based, gradient-based, and spatiotemporal energy-based approaches.

Step 1: Recovery of locally available velocity information

a. Recover locally available velocity information in the first representation using each of the three basic approaches. That is, obtain the velocity estimates U_{cc}, U_{cg}, and U_{ce} using Equations 5.3, 8.5, and 8.13, respectively. Similarly, obtain their covariance matrices S_{cc}, S_{cg}, and S_{ce} using Equations 5.4, 8.6, and 8.14, respectively.

b. Fuse the three estimates on the basis of their covariance matrices using the information fusion theorem. Let the fused velocity estimate and its covariance matrix be given by $V_{cf} = (u_{cf}, v_{cf})$ and S_{cf}, respectively.

Step 2: Propagation of velocity information

a. Plot the current estimate of velocity at the neighboring pixels in the uv plane. Quantify the scatter thus obtained by an estimate $\overline{U}^k$ and a covariance matrix S_n^k.

b. Update the velocity at the pixel under consideration using the following equations (same as Equation 5.15 with U_{cc} and S_{cc} replaced by U_{cf} and S_{cf}, respectively):

$$U^{k+1} = [(S_{cf})^{-1} + (S_n^k)^{-1}]^{-1} [(S_{cf})^{-1}U_{cf} + (S_n^k)^{-1}\overline{U}^k]$$
$$U^0 = U_{cf} \quad \textbf{(9.35)}$$

c. Repeat propagation steps a and b until the change in the flow field over two successive iterations is less than some threshold.

d. Compute the confidence measures associated with the final estimate of velocity as the eigenvalues of the matrix given by $(S_{cf})^{-1} + (S_n)^{-1}$. Use the value of S_n from the last iteration.

Algorithm 4. This algorithm is based on the second representation of velocity: two mutually orthogonal linear constraints on velocity, each with a confidence measure. In essence, this algorithm is an extension of Scott's algorithm [135]. Instead of using only two conservation constraints as discussed in Scott's algorithm, I use six conservation constraints, two each from the three basic approaches. (In another publication [159], I reported an earlier version of this algorithm, which integrated the correlation-based and gradient-based approaches.)

The underlying idea of Algorithm 4 can be explained as follows. Each of the three approaches provides two linear (conservation) constraints on the velocity of the given pixel. Furthermore, the distribution of neighborhood velocities at the current iteration provides two additional linear (neighborhood) constraints on velocity. The updated velocity for the current iteration can be computed as a point in the uv plane that satisfies each of the eight constraints to a degree suggested by its confidence measure. In other words, a pseudo-intersection of the eight constraints weighted by confidence measures gives the updated velocity at the pixel under consideration. This is shown in Figure 9.2. A step-by-step procedure is given below.

Step 1: Recovery of locally available velocity information. Compute the conservation constraints and their confidence measures using each of the three approaches according to Equations 5.5, 8.7, and 8.15, respectively.

Step 2: Propagation of velocity information

a. Compute the neighborhood constraints and their confidence measures as explained in Chapter 5.

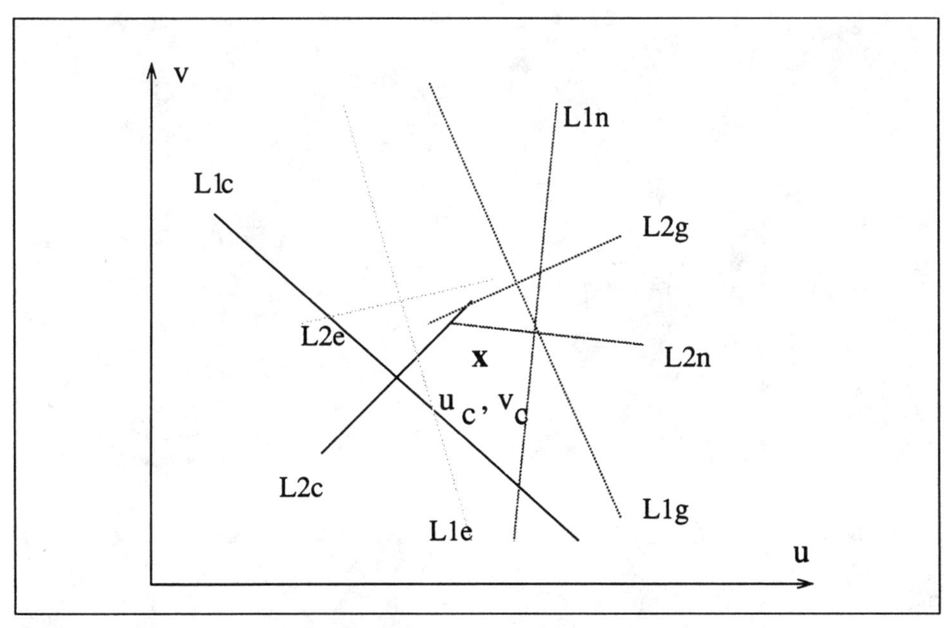

Figure 9.2. Algorithm 4: A graphical interpretation.

b. Update the velocity for the pixel under consideration by minimizing the following error function (sum of the squares of the perpendicular distances from each constraint, weighted by the appropriate confidence measure). Various symbols used in this function have been defined earlier.

$$\mathcal{F} = \mathcal{L}_{1c}^2 C_{1c} + \mathcal{L}_{2c}^2 C_{2c} + \mathcal{L}_{1g}^2 C_{1g} + \mathcal{L}_{2g}^2 C_{2g} + \mathcal{L}_{1e}^2 C_{1e} + \mathcal{L}_{2e}^2 C_{2e} + \mathcal{L}_{1n}^2 C_{1n} + \mathcal{L}_{2n}^2 C_{2n} \tag{9.36}$$

The updated velocity, obtained using a calculus of variations [146], is given by

$$u = \frac{t*r - q*s}{p*q - r^2}$$
$$v = \frac{s*r - t*p}{p*q - r^2} \tag{9.37}$$

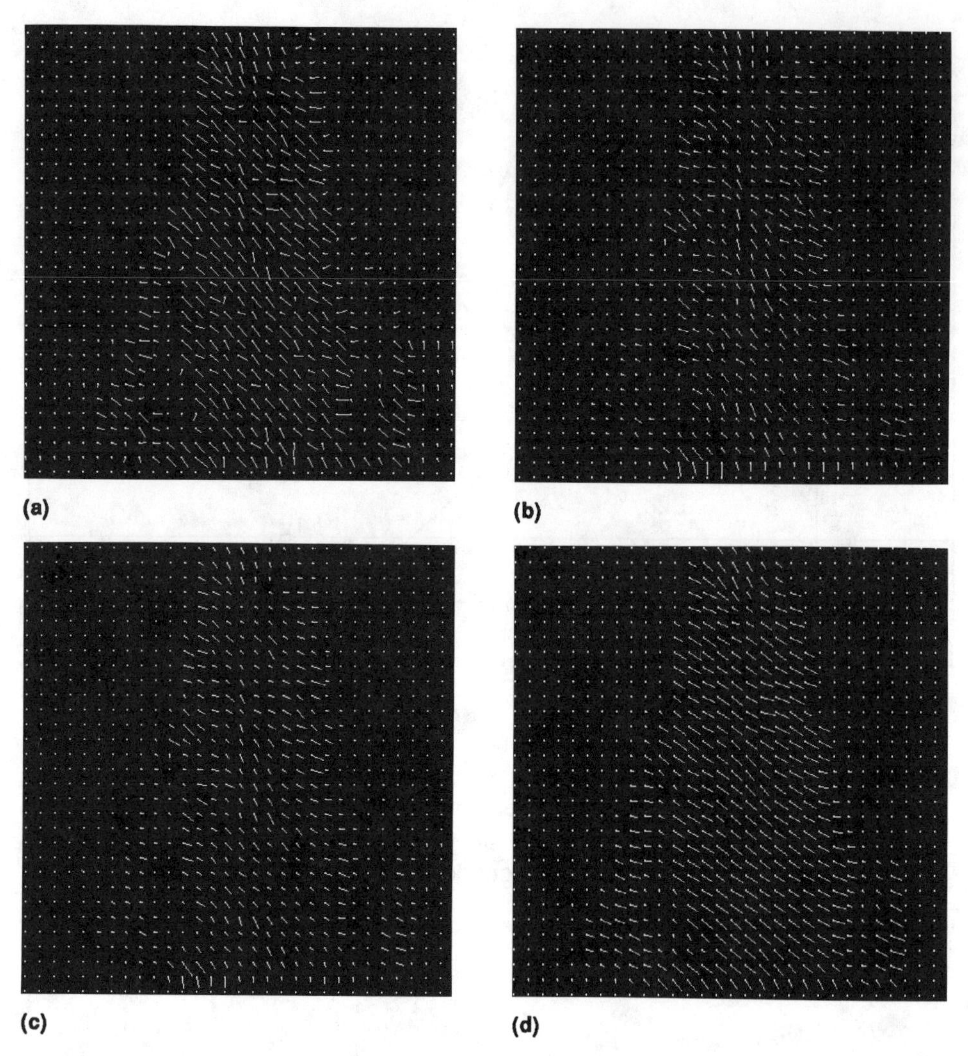

Figure 9.3. Results of applying the integrated scheme on the baby sequence: (a) initial estimates obtained from the correlation-based approach, (b) initial estimates obtained from the gradient-based approach, (c) initial estimates obtained from the energy-based approach, (d) final estimates obtained after fusion and velocity propagation.

where

$$p = C_{1c}\, a_{1c}^2 + C_{2c}\, a_{2c}^2 + C_{1g}\, a_{1g}^2 + C_{2g}\, a_{2g}^2 + \\ C_{1e}\, a_{1e}^2 + C_{2e}\, a_{2e}^2 + C_{1n}\, a_{1n}^2 + C_{2n}\, a_{2n}^2$$

$$q = C_{1c}\, b_{1c}^2 + C_{2c}\, b_{2c}^2 + C_{1g}\, b_{1g}^2 + C_{2g}\, b_{2g}^2 + \\ C_{1e}\, b_{1e}^2 + C_{2e}\, b_{2e}^2 + C_{1n}\, b_{1n}^2 + C_{2n}\, b_{2n}^2)$$

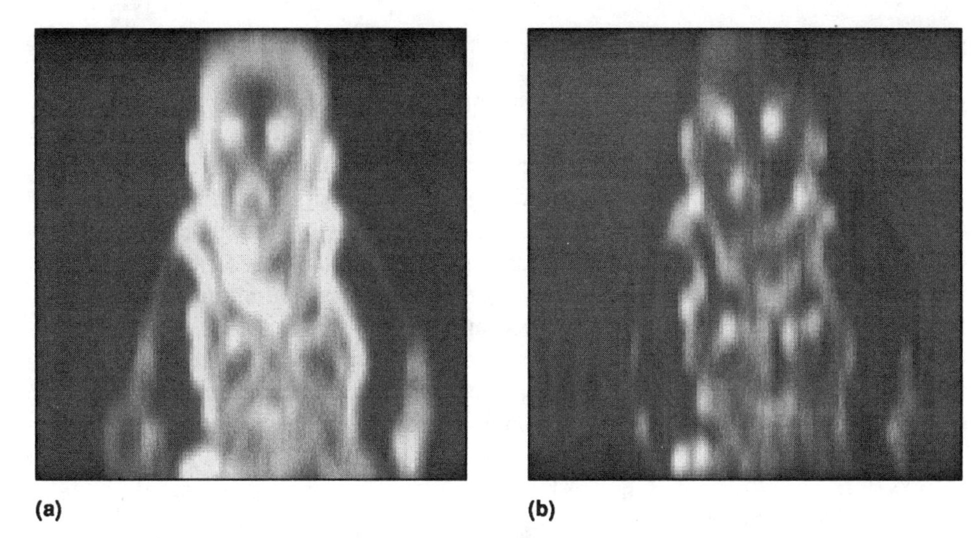
(a) (b)

Figure 9.4. Results of applying the integrated scheme on the baby sequence: (a) maximal and (b) minimal confidence measures associated with the final estimates of velocity.

$$
\begin{aligned}
r &= C_{1c}\, a_{1c}\, b_{1c} + C_{2c}\, a_{2c}\, b_{2c} + C_{1g}\, a_{1g}\, b_{1g} + C_{2g}\, a_{2g}\, b_{2g} + \\
&\quad C_{1e}\, a_{1e}\, b_{1e} + C_{2e}\, a_{2e}\, b_{2e}) + C_{1n}\, a_{1n}\, b_{1n} + C_{2n}\, a_{2n}\, b_{2n}) \\
s &= C_{1c}\, a_{1c}\, c_{1c} + C_{2c}\, a_{2c}\, c_{2c} + C_{1g}\, a_{1g}\, c_{1g} + C_{2g}\, a_{2g}\, c_{2g} + \\
&\quad C_{1e}\, a_{1e}\, c_{1e} + C_{2e}\, a_{2e}\, c_{2e}) + C_{1n}\, a_{1n}\, c_{1n} + C_{2n}\, a_{2n}\, c_{2n}) \\
t &= C_{1c}\, b_{1c}\, c_{1c} + C_{2c}\, b_{2c}\, c_{2c} + C_{1g}\, b_{1g}\, c_{1g} + C_{2g}\, b_{2g}\, c_{2g} + \\
&\quad C_{1e}\, b_{1e}\, c_{1e} + C_{2e}\, b_{2e}\, c_{2e}) + C_{1n}\, b_{1n}\, c_{1n} + C_{2n}\, b_{2n}\, c_{2n})
\end{aligned}
\qquad (9.38)
$$

c. Repeat propagation steps a and b until the change in the flow field over two successive iterations is less than some threshold.

An experiment

I applied Algorithm 3 to the "baby-sequence" of the running example. As mentioned earlier, the correct velocity at each pixel lying on the baby is about one and a half pixels per frame, pointing downward and to the right. The ground-truth optic flow, however, is unknown. In this implementation of the algorithm, I used the same values of various parameters as those used in the implementations shown in Chapters 5 and 8. The initial estimates of optic flow obtained from the correlation-based approach, the gradient-based approach, and the energy-based approach are shown in Figures 9.3a, 9.3b, and 9.3c, respectively. Figure 9.3d shows the final flow field after fusion and propagation. The velocity-propagation procedure was stopped when the magnitude of each component of velocity, rounded off to the second decimal place, did not change. Ten iterations were sufficient for this sequence. Figures 9.4a and 9.4b show the confidence measures associated with the final estimate of velocity.

Conclusion

In this chapter, I have discussed the issue of integration of multiple approaches for estimating optic flow. I have also shown two functionally equivalent algorithms for optic-flow estimation. These algorithms integrate the three approaches for local recovery of optic flow and carry out velocity propagation. In the first algorithm, the process of integration is explicit. It follows directly from the information fusion theorem shown in the previous chapter. In the second algorithm, however, the process of integration is implicit in velocity propagation. In each algorithm, conservation information is computed only once at the onset. Neighborhood information is computed at each iteration and combined with conservation information to give an updated estimate of velocity. I applied this procedure to the "baby-sequence" of the running example. I defer a systematic study of various implementation issues and experimental verification of the overall advantage of integration to future research.

Chapter 10

Conclusion

To keep with the orthodoxy of technical writing, I have titled this chapter "Conclusion," even though the objective of this chapter is not to conclude, but to set the stage for continuation of this work. Understanding optic flow is an endeavor still in its infancy, and an attempt to close the case by enumerating firm conclusions would be a disservice. Such is not my intention. Instead, I briefly summarize the important contributions of this work. Then I discuss the issues that do belong in this book, but have been addressed only cursorily or have not been addressed at all. In doing so, I try to outline the directions for future research, along with possible strategies to pursue them.

Summary of contributions

A significant contribution of this book is an estimation-theoretic framework to compute optic flow. Digitized images of the real world are never noise free. Anything, including optic flow, that must be recovered from noisy images cannot be exact—it must have uncertainty associated with it. The framework proposed in this book formalizes the notion of uncertainty. It does so by posing the problem of optic-flow recovery as that of parameter estimation, where the estimated parameter—optic flow—is in the form of an estimate accompanied by a covariance matrix for each point in the image. The estimation process explicitly attempts to minimize the mean-squared error in the estimated optic flow. The formulation of mean-squared error incorporates

- *conservation error*, which depends on the nature of local spatial variations in the conserved image property, and
- *neighborhood error*, which depends on the nature of local spatial variations in the velocity field.

These two errors originate, respectively, during the two functional steps in optic-flow computation: local recovery of velocity information and propagation of velocity information.

I have shown a detailed implementation of this framework using the correlation-based approach. I have also used this implementation to recover optic flow from a variety of image sequences. The quantitative accuracy of optic flow thus recovered is quite good and compares well with that attainable by, for instance, Anandan's algorithm (which has been shown to provide some of the best published results on real-image sequences). In addition, this framework, as compared with the smoothing-based frameworks described in the past research, does a much better job of preserving the step discontinuities in the flow field during velocity propagation. This is particularly noticeable if there is no texture in the vicinity of the discontinuity. This advantage comes from the current formulation of neighborhood error, which explicitly accounts for the nature of local spatial variations in the flow field. Blurring of flow fields at discontinuities has been a major drawback of several frameworks proposed in the past research. Further, my framework has an explicit notion of initial (prepropagation) and final (postpropagation) confidence measures. The confidence measures, although simply the eigenvalues of the corresponding covariance matrices, are much easier to visualize, and they exhibit a clear relationship with the underlying scene geometry.

This framework lends itself very naturally to applications such as motion-compensated image-sequence enhancement, recursive estimation of 3D scene geometry from optic flow using Kalman-filtering-based techniques, and so on. I have shown some experimental results of such applications. Motion-compensated enhancement of image sequences is essentially a 2D application where the optic-flow field is used only to warp images. The accuracy of the optic-flow fields recovered by my framework appears sufficient for such a task. Estimation of scene geometry, on the other hand, is a 3D application. Here, the optic-flow field is used to invert the perspective projection to recover scene depth. Even though a theoretical solution to this problem exists, it is, in practice, quite susceptible to noise. A 1 percent error in the flow field, could produce, for instance, a 100 percent error in the scene depth. While the two experiments shown in Chapter 7 can recover scene depth with very good accuracy, they represent an exception, rather than a rule. I attribute their accuracy to highly controlled experimental conditions (including very accurate knowledge of camera motion).

The second major contribution of this book is in the context of unification. There are two aspects of unification: one pertaining to conservation information and the other to neighborhood information. With respect to conservation information, I have shown that the estimation-theoretic framework applies identically to each of the three basic approaches. That is, the same overall computational scheme can be used irrespective of how conservation information is recovered, as long as it is recovered in the form that can be used by the framework: a response distribution in *uv* space, characterized by an estimate and a covariance matrix.

I have shown that conservation information can, in fact, be recovered in this form by any of the three approaches, using a single procedure. This procedure is composed of three essential steps:

1. Select the invariant and derive it from intensity imagery.
2. Compute error distribution and response distribution.
3. Interpret response distribution to get an estimate of velocity along with its covariance matrix.

With respect to neighborhood information, I have shown that the formulation of neighborhood error used in this framework is quite general. It reduces to the "smoothness error" used in the past frameworks based on conventional smoothing [34, 80] under specific operating conditions. In other words, the velocity-propagation procedure behaves in the same way as some of the smoothing-based procedures with certain specific types of imagery.

I have also shown that this framework can be used to integrate multiple approaches for recovering optic flow. I showed one implementation of the integrated scheme and applied it to recover optic flow from the "baby sequence" of the running example. I have not, however, pursued either a detailed analysis of implementation issues or an experimental investigation of the overall advantage of integration.

Directions for future research

There are several areas—including those discussed in the previous section—that require further research. In this section, I enumerate the important issues, along with some possible strategies to investigate them.

Conservation information. The behavior of response distribution needs to be analyzed in greater detail, irrespective of the approach used to recover it. The current version of the framework gives a good initial estimate of velocity only if the response distribution is unimodal. If the response distribution is multimodal, it gives an incorrect estimate. It does, however, associate a low confidence with the estimate—a feature that is very desirable.

Multimodality typically arises in textured regions, occlusions, or transparent moving objects. Even though using multiple frames (in the correlation-based approach) serves as a useful heuristic in some of these regions, the problem of multimodality cannot be eliminated altogether. In this case, taking a weighted least-squares estimate based on the complete distribution is obviously incorrect. It would be desirable to have a mechanism to identify the correct peak in the response distribution and use it to estimate velocity. Further research needs to be done to formalize the notion of "correct" peak. Alternatively, if the two peaks are not equally strong, we might consider using the median (instead of the mean) as the estimate of velocity.

As mentioned in Chapter 5, there are inevitable deviations from the assumption of conservation of image properties over time (because of photometric changes, perspective distortions, rotation, and so on). Further research needs to be done to determine the overall effect of these deviations on the estimates of optic flow. Also, the possibility of dealing with these deviations needs to be explored.

Neighborhood information. In the current version of the framework, the velocity-propagation procedure uses only the estimate of velocity at neighboring pixels. It does not use the covariance matrix associated with the estimate. It is plausible that the knowledge of the covariance matrix might assist in identifying motion discontinuities, thus making the velocity-propagation procedure even more robust at discontinuities. More research needs to be done to develop a mechanism that allows a natural inclusion of covariance information in the formulation of neighborhood error.

Overall framework. There are several issues concerning the overall framework that need further investigation.

Incorporating the temporal dimension. The current version of the framework works on the "batch-processing" paradigm. All the images are fed to the algorithm at once, and a flow field is produced at the end of the computation. By definition, optic flow is an instantaneous, time-varying phenomenon. As such, extending the current framework to incorporate the temporal dimension would be useful and efficient in an application where the flow field must be updated as time progresses. Since the framework is inherently estimation-theoretic, it lends itself easily to recursive update procedures—such as Kalman filtering—for incremental estimation of optic flow. I discuss some preliminary efforts in this direction in Appendix A.

Investigating the effects of error dependence. The formulation of the optimization problem assumes that conservation error and neighborhood error are independent. In the current implementation, however, neighborhood information is derived from conservation information. This makes the two errors dependent. In the experiments conducted in this book, the dependence does not seem to corrupt the estimated optic flow noticeably. However, an investigation of the effects of this dependence would certainly be very useful in predicting the performance of the framework with any given imagery.

Making the errors independent. With respect to the problem mentioned above, efforts could be made to ensure that the two errors are, in fact, independent. A possible technique would be to derive conservation and neighborhood information from two different approaches. For example, conservation information could be derived using the correlation-based approach, whereas neighborhood information could be derived from the velocity estimates obtained using the gradient-based approach. For this strategy to work, we must assume that the velocity estimates obtained from the correlation-based and gradient-based approaches are independent. This assumption itself needs to be verified, not only in the current context, but also for the integration of multiple approaches to be useful.

Establishing a good coarse-to-fine control strategy. In the current hierarchical implementation of the framework, the control strategy to propagate velocity estimates at a coarse resolution to a finer resolution is very simplistic. A good coarse-to-fine control strategy needs to be established to deal with motion discontinuities coherently in a hierarchical implementation. It is not clear how motion discontinuities at the highest level of resolution appear at lower resolutions, for instance, in Burt's band-pass pyramid.

Verifying quantitative correctness. For imagery containing motion discontinuities, the quantitative correctness of the recovered flow fields must be verified. In this book, I have done such testing primarily for imagery without motion discontinuities. For imagery with discontinuities, I have done quantitative verification only cursorily, relying mostly on qualitative verification.

Integration of multiple approaches. The issue of integration has been discussed very briefly in this book. Chapter 9 lists several topics that need to be addressed. Summarizing, they are

- verification of the assumption of independence of conservation information obtained from multiple approaches,
- investigation of the overall advantage of integration, and
- study of various implementation issues, such as selection of normalization parameters.

Appendix A

Incremental Estimation of Optic Flow Using a Kalman Filter

Many real-time applications of optic flow, such as navigation and tracking, require on-line, incremental recovery of optic flow, along with an associated measure of uncertainty. This appendix describes a framework for incremental estimation of optic flow. In this framework, the process of updating velocity estimates is extended to use spatial *as well as* temporal neighborhoods. Kalman filtering is used as the basis for the update. After I give an overview of the framework, I discuss computational aspects of its various components and describe the results of applying it to a variety of image sequences.

Overview

Kalman filtering provides a convenient platform for incremental estimation (of a quantity) in a setting where multiple measurements (that relate to the quantity) are acquired over time. Figure A.1 shows a schematic diagram of the Kalman filter, and Figure A.2 summarizes its associated equations. The figures show that the filter is based on a linear *measurement model* and operates in two phases—*prediction phase* and *update phase*—to obtain an unbiased estimate with a minimum mean-squared error [158]. Below, I give a brief description of the underlying model and the two phases.

The measurement model can be summarized as follows. The quantity to be estimated is the state vector U_t, whose evolution over time is specified through the state-transition matrix Φ_{t-1} and the addition of Gaussian *process* noise with a covariance Q_{t-1}, as shown in Equation A.1 (in Figure A.2). Further, the state vector U_t is related linearly to the measurement vector D_t through the measurement matrix H_t and the addition of Gaussian *sensor* noise with a covariance R_t, as shown in Equation A.2. We assume that the process noise and sensor noise are uncorrelated. In the prediction phase, the previous state estimate $\hat{U}^+_{t-1}$ and its covariance P^+_{t-1} are extrapolated to the predicted state vector $\hat{U}^-_t$ and its covariance P^-_t. The predicted covariance is used to compute the Kalman gain K_t. In the update phase, the measurement residual $D_t - H_t\hat{U}^-_t$ is weighted by the Kalman gain K_t and added to the predicted state $\hat{U}^-_t$ to yield the updated state $\hat{U}^+_t$.

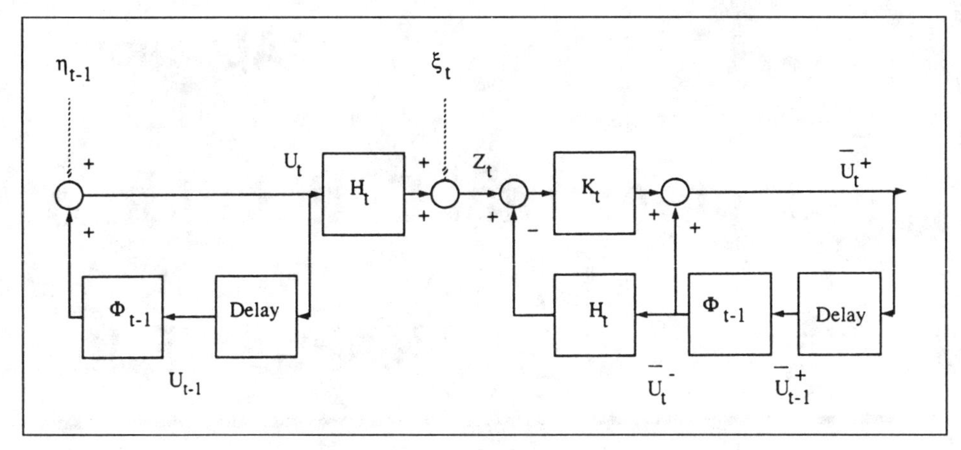

Figure A.1. A schematic diagram of the Kalman filter.

Measurement model

$$U_t = \Phi_t U_{t-1} + \eta_{t-1}, \quad \eta_t \sim N(0, Q_t) \tag{A.1}$$

$$D_t = H_t U_t + \xi_t, \quad \xi_t \sim N(0, R_t) \tag{A.2}$$

$$E[\eta_t \xi_t^T] = 0 \tag{A.3}$$

Prediction phase

$$\hat{U}_t^- = \Phi_{t-1} \hat{U}_{t-1}^+ \tag{A.4}$$

$$P_t^- = \Phi_{t-1} P_{t-1}^+ \Phi_{t-1}^T + Q_{t-1} \tag{A.5}$$

Update phase

$$\hat{U}_t^+ = \hat{U}_t^- + K_t [D_t - H_t \hat{U}_t^-] \tag{A.6}$$

$$P_t^+ = [I - K_t H_t] P_t^- \tag{A.7}$$

$$K_t = P_t^- H_t^T [H_t P_t^- H_t^T + R_t]^{-1} \tag{A.8}$$

Figure A.2. Kalman filter equations.

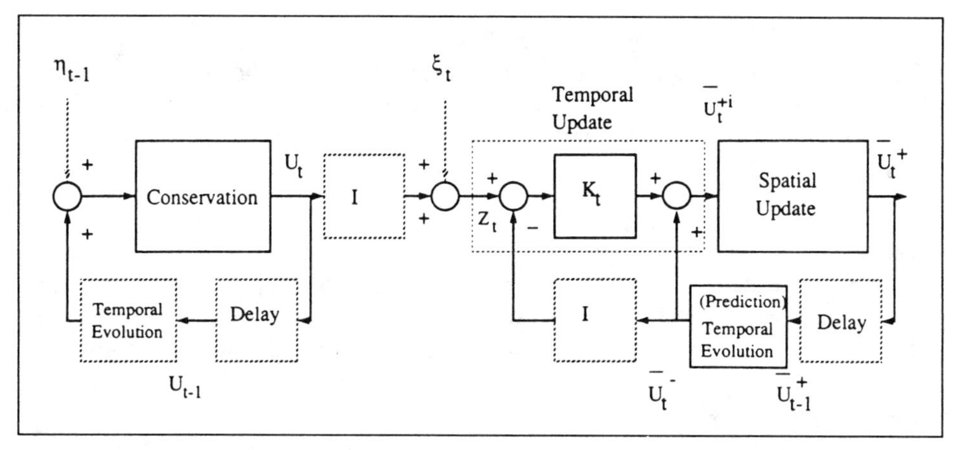

Figure A.3. A scheme for optic-flow estimation embedded in the Kalman filter. The blocks drawn in dotted lines are implicit and do not have to be implemented.

Figure A.3 shows the proposed optic-flow estimation scheme embedded in the Kalman filter discussed above. It comprises the following components (described in detail in the next section):

- *Conservation.* As discussed earlier in the book, conservation information is in the form of an optic-flow vector $U_c = (u_c, v_c)$ for each pixel, accompanied by a covariance matrix S_c. A comparison between Figures A.1 and A.3 reveals that recovering conservation information corresponds to measurement. Therefore, $D_t = U_c$ and $R_t = S_c$. Further, since the objective is to estimate optic flow, the state vector is the optic-flow vector itself. This implies that the matrix H_t is a 2×2 identity matrix.
- *Temporal update.* This amounts to computing the Kalman gain K_t and updating the predicted state estimate $\hat{U}_t^-$ and its covariance P_t^-. I refer to the updated state as $\hat{U}_t^{+i}$ and the updated covariance as P_t^{+i} (the superscript i stands for *intermediate*) because they have to undergo spatial update before the "true" state $\hat{U}_t^+$ and covariance P_t^+ can be obtained.
- *Spatial update.* As discussed earlier, the estimates of optic flow obtained from conservation information can be noisy, especially in the areas of uniform intensity. Spatial update serves to "fill in" these areas. As shown in Chapter 5, this is accomplished by using the velocity information contained in small spatial neighborhoods.
- *Prediction.* This involves implementing Φ_{t-1} on the basis of the knowledge about how optic flow at a point evolves with time. The output of this stage is the state estimate $\hat{U}_t^-$ and its covariance P_t^-.

Computational details

In this section, I describe in detail the proposed optic-flow estimation scheme embedded in the Kalman filter.

Conservation information. I discussed conservation information in detail in Chapters 4, 5, and 8. It gives an estimate of velocity U_c along with a covariance matrix S_c. As discussed earlier, U_c and S_c serve as the parameters D_t and R_t for the measurement model of the Kalman filter.

Temporal update. The objective of this stage is to integrate the new "measurements" (output of the conservation stage) with predicted estimates of optic flow. This is done by computing the Kalman gain K_t and using it to update the state vector U_t (optic flow) and its covariance P_t. From Equation A.8 (in Figure A.2), the Kalman gain is given by

$$K_t = P_t^- H_t^T [H_t P_t^- H_t^T + R_t]^{-1} \tag{A.9}$$

Substituting $H_t = I$ and $R_t = S_c$, we obtain

$$K_t = P_t^- [P_t^- + S_c]^{-1} \tag{A.10}$$

Substituting this in Equation A.6 along with $D_t = U_c$ and $H_t = I$, the updated optic-flow estimate is given by

$$\hat{U}_t^{+i} = \hat{U}_t^- + P_t^- [P_t^- + S_c]^{-1} [U_c - \hat{U}_t^-] \tag{A.11}$$

and the covariance associated with this estimate is given by

$$P_t^{+i} = [(P_t^-)^{-1} + (S_c)^{-1}]^{-1} \tag{A.12}$$

Spatial update. Spatial update simply amounts to incorporating the spatial neighborhood information (discussed in Chapters 4 and 5) to propagate velocity from regions of low uncertainty (or conversely, high confidence), such as corners and textured regions, to those of high uncertainty, such as edges and flat regions. Spatial update can be viewed as a part of Kalman filtering that incorporates prior knowledge about the nature of spatial variation (or smoothness) of the flow field.

As discussed in Chapter 5, neighborhood information gives an estimate of velocity $\bar{U}$ along with a covariance matrix S_n. In the Kalman formalism, this information is used as "prior" knowledge (given that sufficient time has elapsed since the estimation process began) that can be fused with the newly acquired knowledge (the estimate $\hat{U}_t^{+i}$ and the covariance matrix P_t^{+i}). The updated estimate $\hat{U}_t^+$ that weights the two sources of information appropriately to minimize the squared error is given by [144]

$$(P_t^{+i})^{-1}(\hat{U}_t^{+} - \hat{U}_t^{+i}) + S_n^{-1}(\hat{U}_t^{+} - \overline{U}) = 0 \quad \textbf{(A.13)}$$

On the basis of the arguments given in Chapter 5, an iterative solution for the current velocity estimate can be written as [147]

$$(\hat{U}_t^{+})^{k+1} = \left[(P_t^{+i})^{-1} + S_n^{-1}\right]^{-1}\left[(P_t^{+i})^{-1}\hat{U}_t^{+i} + S_n^{-1}\overline{U}^{k}\right]$$

$$(\hat{U}_t^{+})^{0} = \hat{U}_t^{+i} \quad \textbf{(A.14)}$$

and the covariance matrix P_t^{+} associated with the current estimate of velocity is given by

$$\left[(P_t^{+i})^{-1} + S_n^{-1}\right]^{-1}$$

where S_n^{-1} is computed from the previous iteration. The eigenvalues of this matrix depict the confidence measures corresponding to the current estimate. In the experiments reported here, only one iteration of spatial update is performed per frame. More iterations of spatial update would reduce the number of frames required to reach an acceptable level of accuracy. This would, however, require that more time be allowed between successive frames.

Prediction. The prediction stage must predict optic flow as well as its covariance for each pixel in the current image. In essence, prediction amounts to defining Φ, the way in which optic flow, that is, the state vector, evolves over time. My definition of Φ is based on the assumption of temporal coherence. That is, the velocity at a point "does not change drastically" over small periods of time. The prediction scheme discussed below is based on the one used by Matthies, Szeliski, and Kanade for disparity extrapolation [117].

Scene motion shifts the point X_{t-1} in the previous image to the point X_t in the current image. We assume that $\hat{U}_t^{-}$ at point X_t is the same as the velocity $\hat{U}_{t-1}^{+}$ at point X_{t-1}. Therefore, Φ is simply the function that warps X_{t-1} to X_t. In general, this warping function will predict optic flow at locations that do not coincide with pixel locations in the new image. This is because the optic-flow vectors in the previous image do not necessarily have both components as exact integers. Therefore, we need to resample the predicted optic-flow field to obtain the estimates at pixel locations. For this purpose, we form a unit square around each pixel in the current image. We take the correct predicted estimate at this pixel to be a bilinear interpolation of the velocities of warped pixels that lie within this square.

The uncertainty of optic flow will increase during prediction because of inaccuracies in modeling temporal coherence, camera optics, and so on. We account for this with a simple multiplicative amplification of covariance

$$P_t^- = (1 + \varepsilon)\, P_{t-1}^+ \tag{A.15}$$

I have chosen a value of 0.05 for the parameter ε. The amplified covariance is warped and interpolated in exactly the same fashion as the optic flow itself.

Experiments

Keeping with the spirit of Chapter 5, I describe one experiment from each of the two categories: quantitative and qualitative.

Quantitative experiment. The imagery for this experiment corresponds to the textured poster used in Chapter 5. Using a configuration similar to the one used earlier, 12 frames are shot at regular intervals as the camera translates horizontally by 0.6 inches, in a plane perpendicular to its optical axis. Between the first two frames, the image displacement is roughly four pixels where the poster is closest to the camera and roughly two pixels where the poster is the farthest from the camera. The corresponding displacements between the last two frames are six and three pixels, respectively. The exact amount of camera translation as well as the distance of the lens from the rigid mount is recorded. The camera is then calibrated and its focal length determined. The "correct" flow field for each pair of frames is determined using the theory developed by Waxman and Wohn [102]. The images are low-pass filtered and subsampled to get a resolution of 128×128 using Burt's technique [136]. Both components of image velocity at each point are divided by four to get the correct flow field corresponding to the reduced image size.

Four frames of the sequence (first, fourth, eighth, and 12th) are shown in Figures A.4a, A4.b, A.4c, and A.4d, respectively. Figures A.5a and A.5b show the two confidence measures (inverses of the eigenvalues of P_t^+) after the first two frames have been processed (at $t = 2$). As expected, the regions where the first confidence measure is high correspond to the edges in the intensity image, and the regions where both confidence measures are high correspond to corners in the intensity image. Figures A.5c and A.5d show the corresponding confidence measures at $t = 12$. It is apparent that, in general, at any given pixel, the confidence grows over time, and as time progresses, confidence "propagates outward" from the regions of high initial confidence. These two effects can be attributed to temporal and spatial update, respectively. Figures A.6 and A.7 show the optic-flow estimate $\hat{U}_t^+$ at $t = 2$ and $t = 12$, respectively. The graph in Figure A.8 tracks the percentage root-mean-squared error in the optic-flow estimates (the error being the magnitude of the vector difference between the ground-truth velocity and the estimated velocity). Computed over the entire image, the root-mean-squared error after 12 frames is 2.7 percent. If, however, we consider only the 25 percent most confident pixels, the root-mean-squared error after 12 frames is only 1.2 percent.

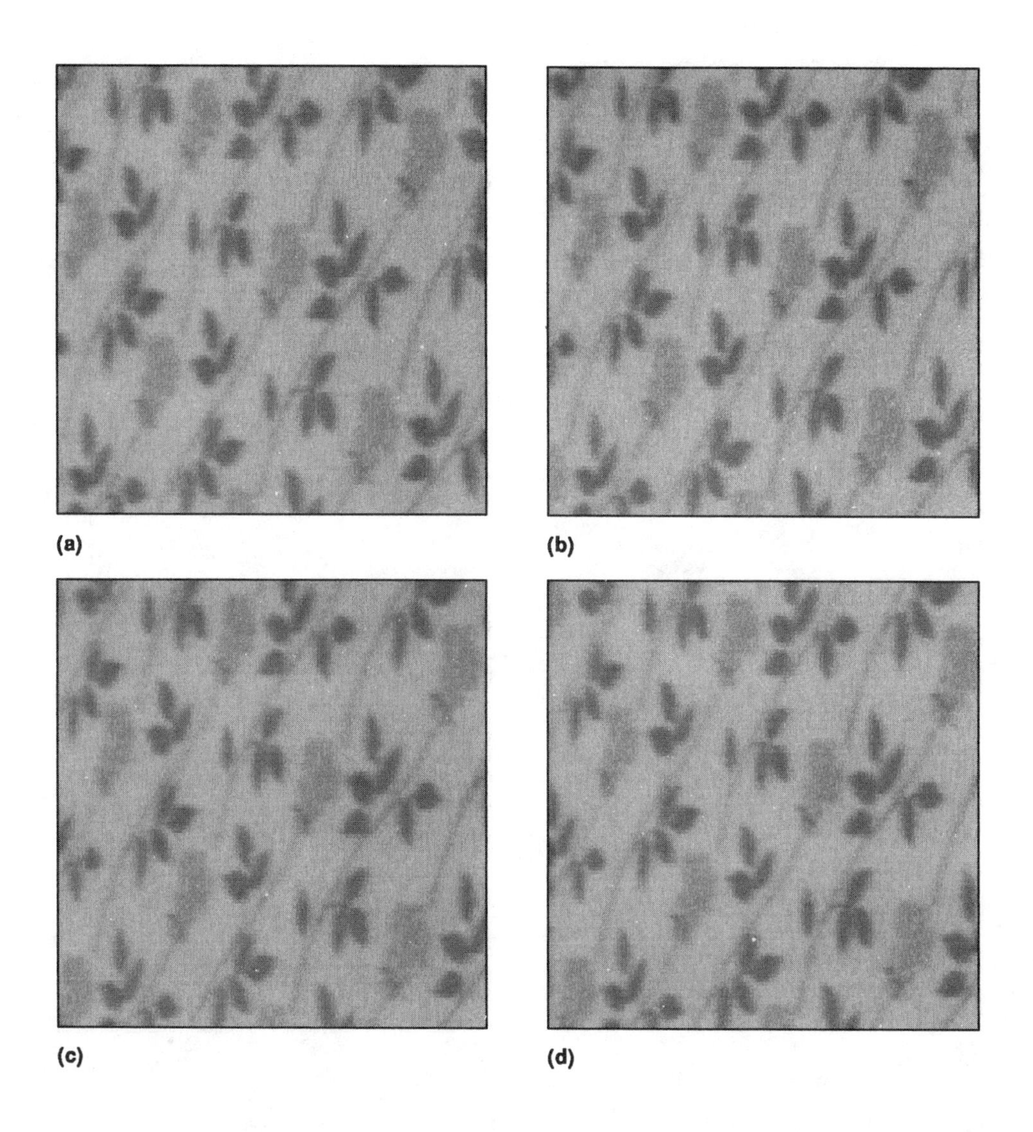

Figure A.4. The poster experiment: (a) first frame, (b) fourth frame, (c) eighth frame, (d) 12th frame.

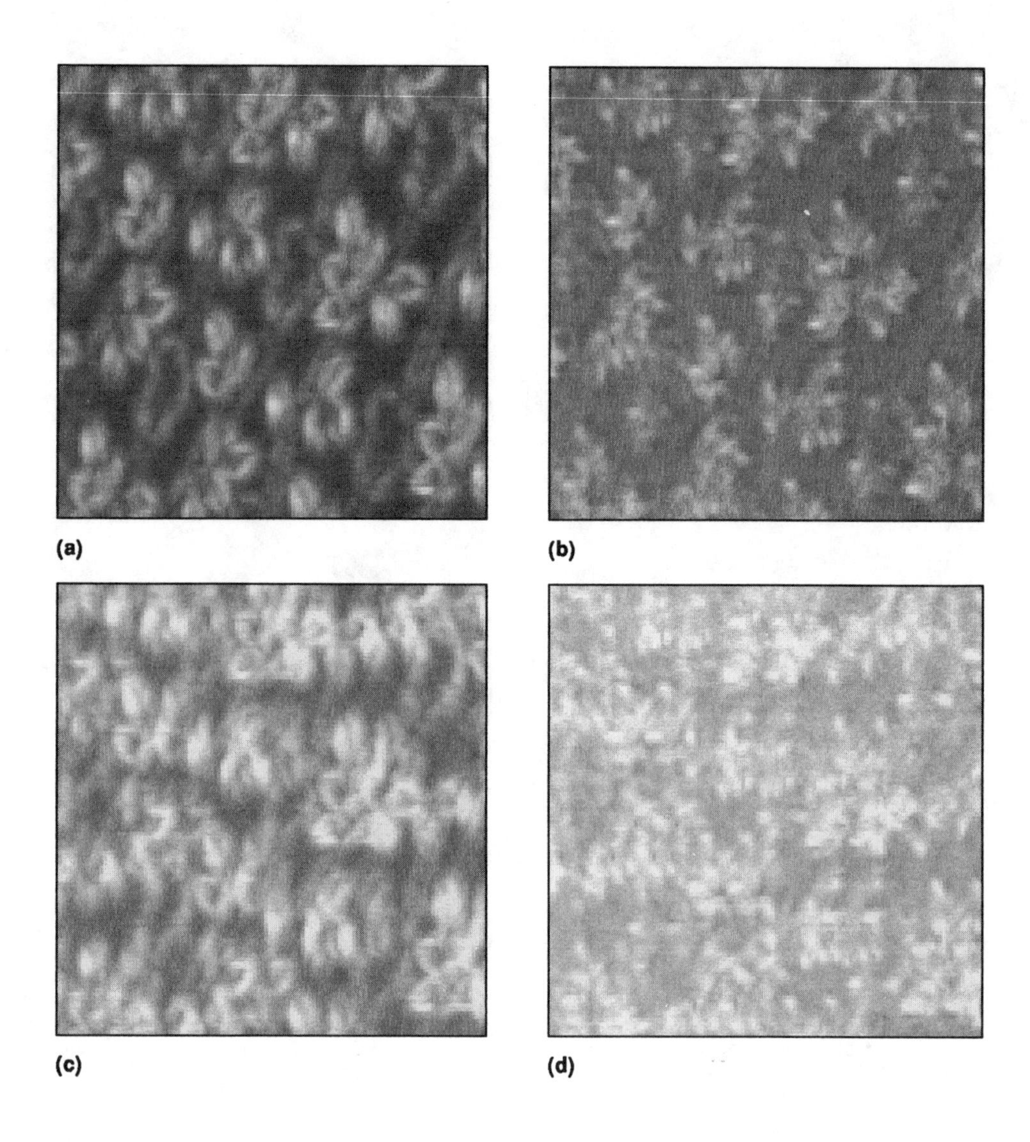

Figure A.5. The poster experiment: (a, b) confidence measures computed as the eigenvalues of the covariance matrix P_t^+ at time $t = 2$, (c, d) the corresponding confidence measures at time $t = 12$.

Figure A.6. The poster experiment: the estimate of optic-flow field $\hat{U}_t^+$ at $t = 2$.

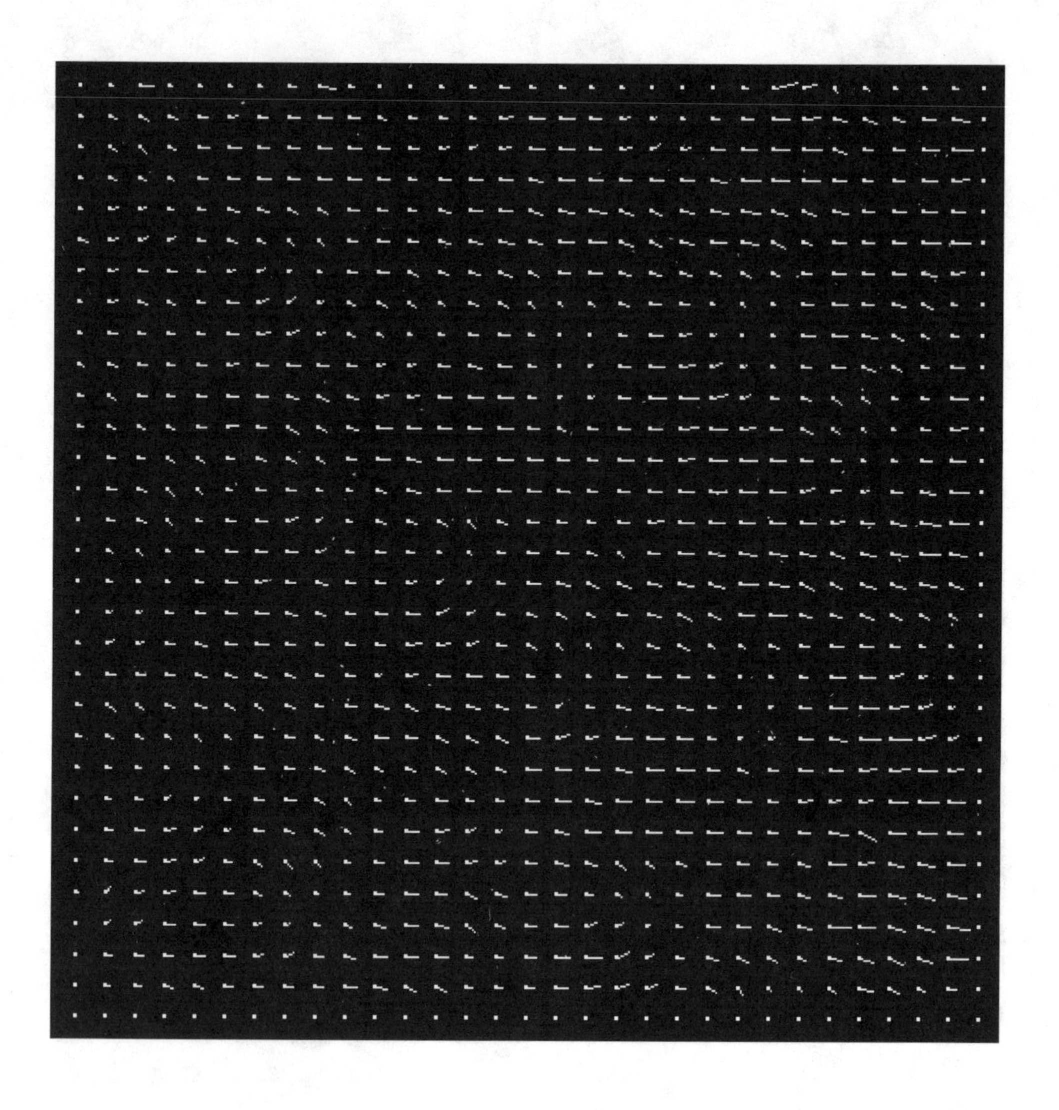

Figure A.7. The poster experiment: the estimate of optic-flow field $\hat{U}_t^+$ at $t = 12$.

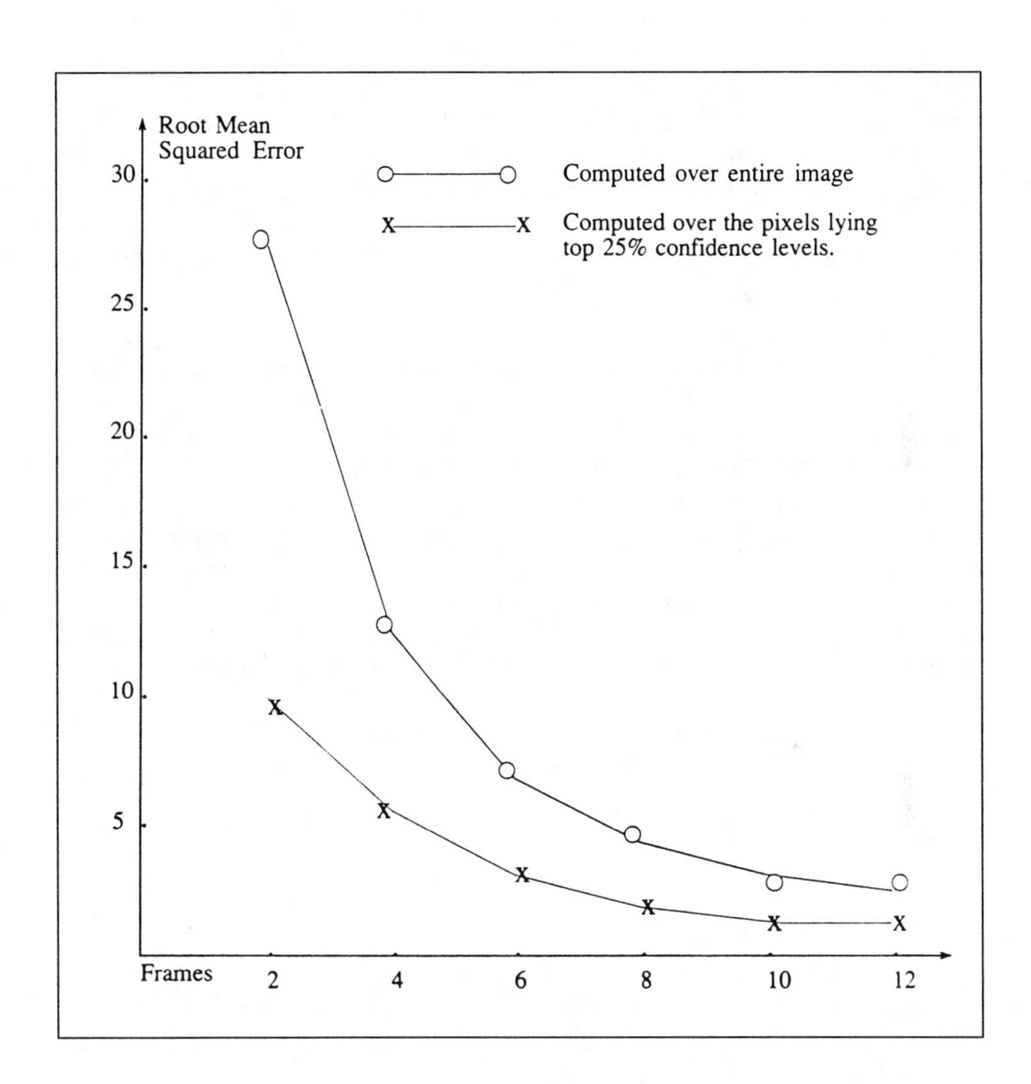

Figure A.8. Root-mean-squared error in velocity estimates as a function of time.

Qualitative experiment. The objective of this experiment is to judge the qualitative correctness of flow fields recovered by the algorithm in realistic environments. The experiment described below uses 12 frames of an image sequence of an outdoor scene (the road scene used in Chapter 7). Figure A.9 shows the first, fourth, eighth, and 12th frames (clockwise from the top-left corner). The principal moving objects in the scene are the two cars (because of the wind, there is slight movement in the trees, too). The ground-truth flow field is not known. The original frames are 256×256 in resolution and are subsampled to 128×128 for computing optic flow.

Figures A.10a and A.10b (top left and right) show the two confidence measures (inverses of the eigenvalues of P_t^+) at $t = 2$. Figures A.10c and A.10d (bottom left and right) show the corresponding confidence measures at $t = 12$. Once again, it is apparent that, in general, at any given pixel, the confidence grows over time, and as time progresses, confidence "propagates outward" from the regions of high initial confidence.

Figures A.11 and A.12 show the optic-flow estimate $\hat{U}_t^+$ at $t = 2$ and $t = 12$, respectively. Clearly, the regions corresponding to the cars show noticeable optic flow. Some of the sky region also shows movement, which is incorrect. I attribute this to "video shatter" (since the sequence was acquired using a camcorder). This region has very low confidence even after sufficient frames have been processed. This is because there is usually very little spatial and temporal coherence in the shatter.

Figure A.13 shows the optic-flow estimates at $t = 12$ at a higher resolution in the regions corresponding to the two cars. The top-left and bottom-left images show the flow field in a square (roughly 20×20 pixels in size) surrounding each car. The top-right and bottom-right images show the same flow fields superimposed on the corresponding regions of the original intensity image. It is noteworthy that the optic flow on the car on the right (bottom images) does not bleed into the thin pole that occludes it.

Figure A.9. The outdoor scene experiment (clockwise from top left): first, fourth, eighth, and 12th frames of the image sequence.

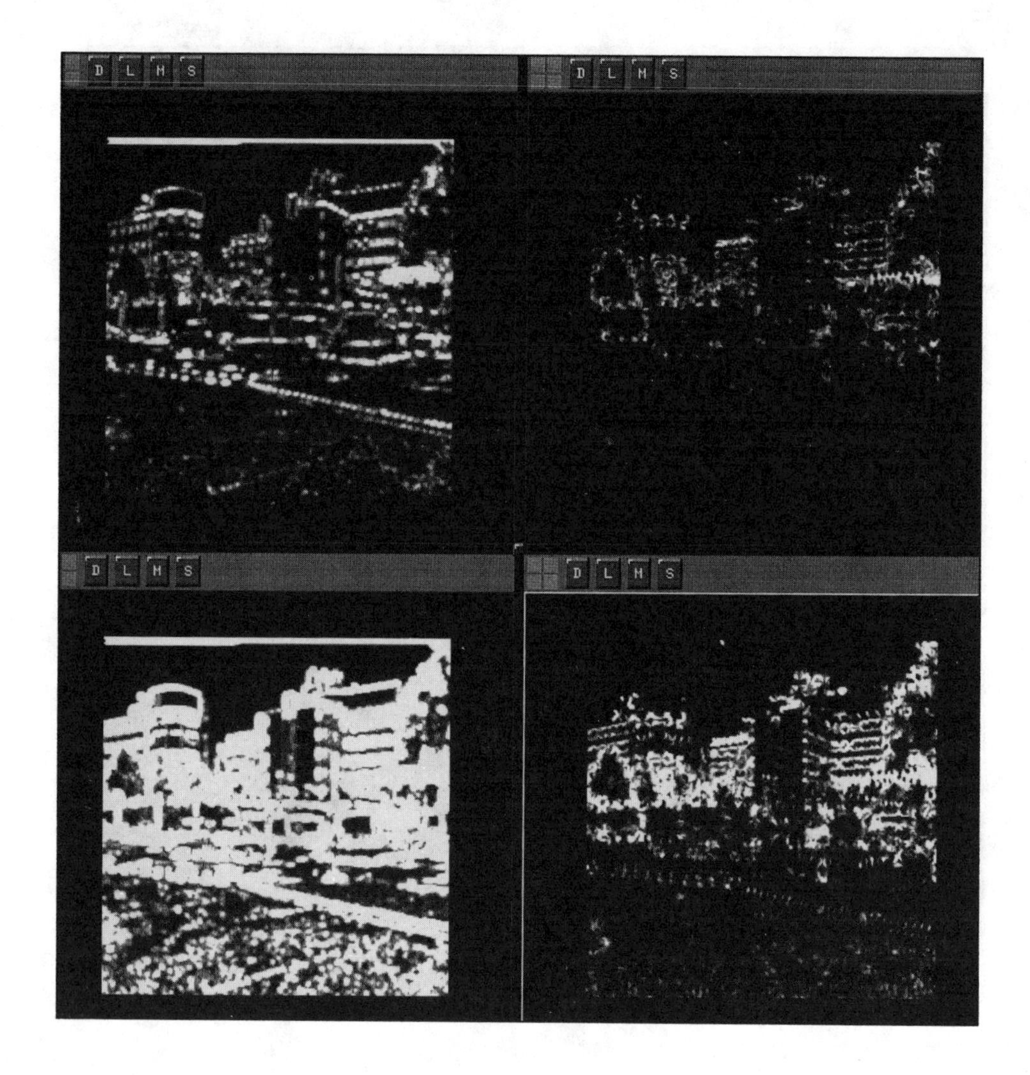

Figure A.10. Top left and right: confidence measures for the outdoor scene experiment computed as the eigenvalues of the covariance matrix P_t^+ at time $t = 2$. Bottom left and right: the corresponding confidence measures at time $t = 12$.

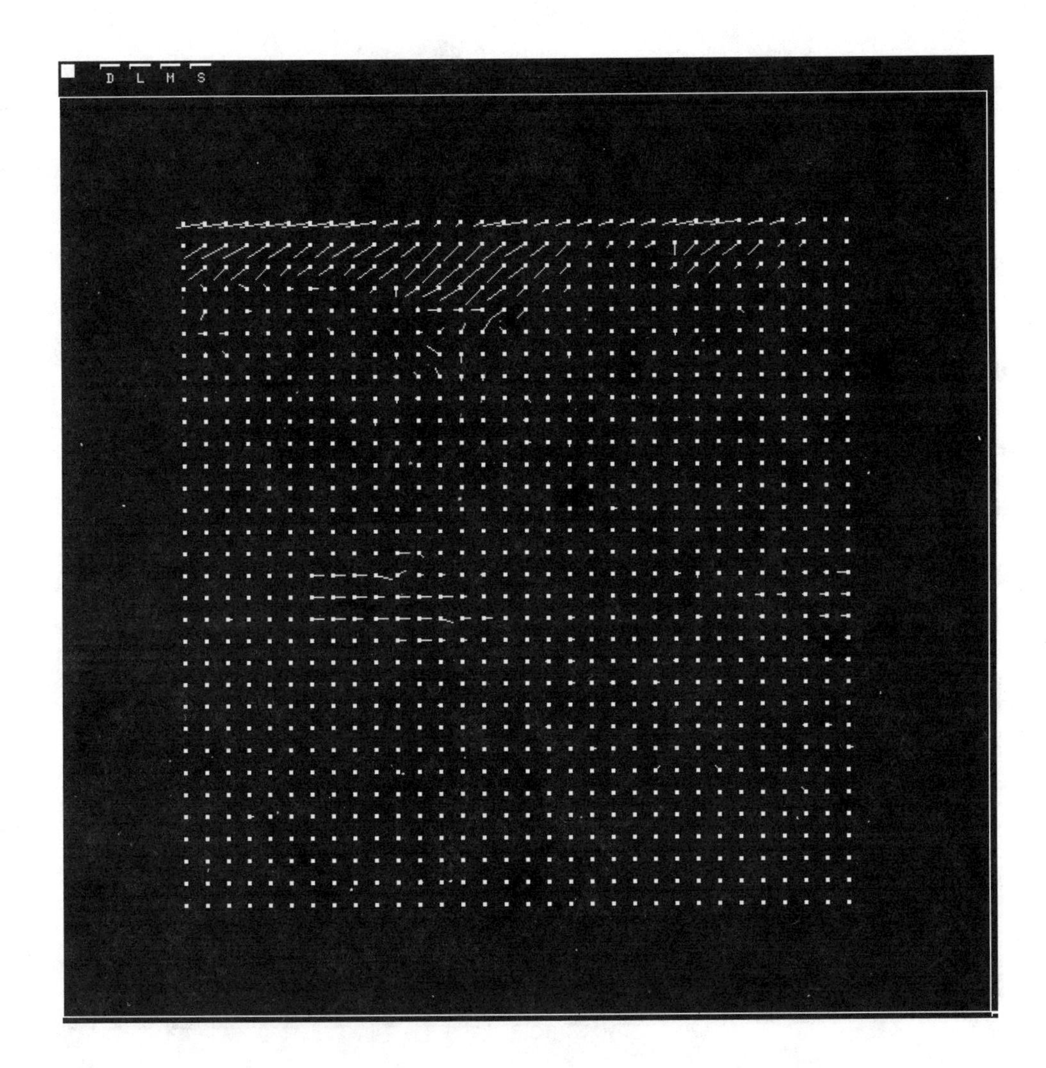

Figure A.11. The outdoor scene experiment: the estimate of optic-flow field $\hat{U}_t^+$ at $t = 2$.

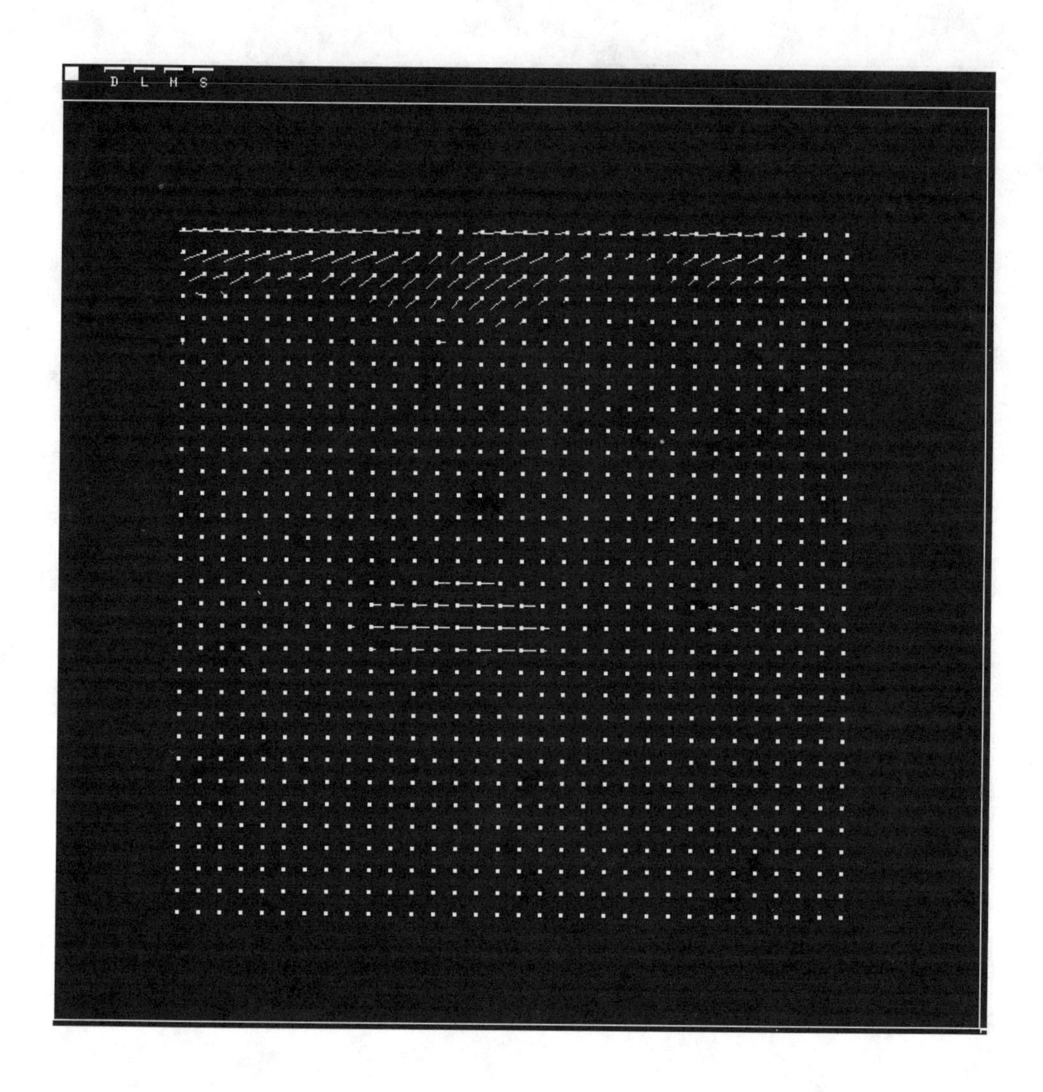

Figure A.12. The outdoor scene experiment: the estimate of optic-flow field $\hat{U}_t^+$ at $t = 12$.

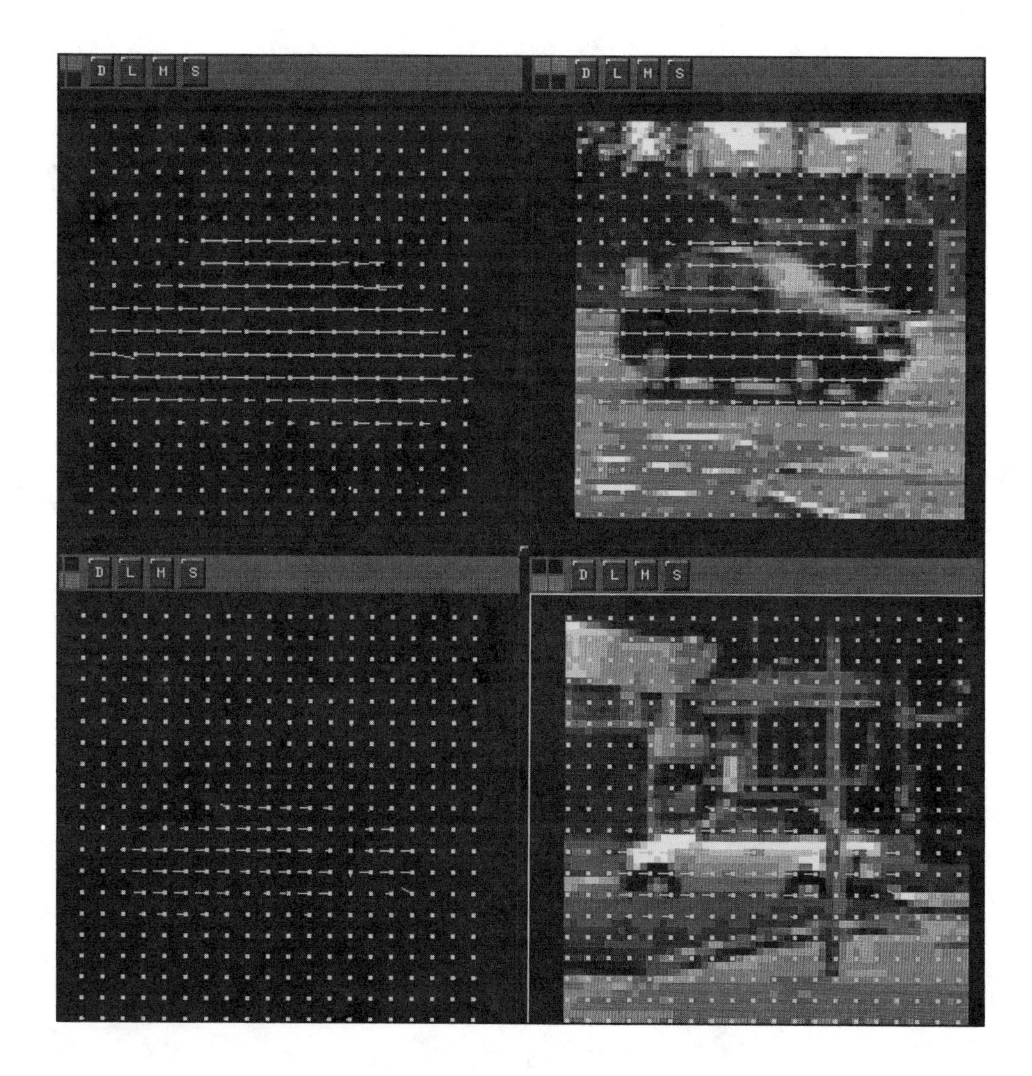

Figure A.13. Top left and bottom left: flow field for the outdoor scene experiment in a small region around each car. Top right and bottom right: the flow field superimposed on the corresponding regions of the intensity image.

Appendix B

Kalman-Filtering-Based Depth Estimation

Matthies, Szeliski, and Kanade [117] reported a Kalman-filtering-based algorithm to recover dense depth maps from optic flow in the case of a stationary scene and known one-dimensional camera motion. This algorithm requires that an estimate of optic flow be produced along with its covariance for each new frame acquired (in a time-sequence) and that it be used to update the existing estimate of disparity (reciprocal of depth) and its variance. The principal advantage of such a scheme is that the uncertainty in depth estimates decreases with time.

Matthies et al. used Anandan's smoothing-based algorithm [80] to estimate optic flow and performed error analysis on the SSD surface to compute its variance. I have adapted their algorithm to use the framework for optic-flow estimation discussed in this book instead of Anandan's algorithm. Because of the new framework's discontinuity-preserving nature, the discontinuities in the depth field are better defined. This makes 3D feature extraction (for interpretation of depth fields) more reliable. Fundamentally, this framework for depth estimation is identical to the one discussed in Appendix A, except that it gives scene depth as its final output (instead of optic flow).

In designing the Kalman filter for depth estimation, Matthies et al. [117] chose the state vector $u(x, y)$ to be the reciprocal of depth—that is, the disparity—linearly related to optic flow. They denoted the variance of disparity by σ_d^2. I use the same representation.

With the scheme shown in Figure A.3, the function of the Conservation block is to produce a measurement of disparity d and its variance σ_d^2. This is accomplished by scaling the estimate of velocity U_c to a disparity measurement by making use of the camera parameters and the known camera motion [117]. Since the velocity is one dimensional, its variance in the direction of motion (one of the diagonal elements of the covariance matrix S_c) gives the variance of disparity by multiplication with the same scale factor. All other computations remain exactly the same as in Appendix A.

References

[1] J.E. Cutting. Four assumptions about invariance in perception. *J. Experimental Psychology: Human Perception and Performance*, 9:310-317, 1983.

[2] H.E. Burton. The optics of Euclid. *J. Optical Soc. America*, 35:357-372, 1945.

[3] H. von Helmholtz. *Physiological Optics*, Vol. 3, third ed., J.P.C. Southall, trans., Optical Soc. America, Menasha, Wis., 1925.

[4] W.R. Miles. Movement in interpretations of the silhouette of a revolving fan. *Am. J. Psychology*, 43:392-404, 1931.

[5] H. Wallach and D.H. O'Connell. The kinetic depth effect. *J. Experimental Psychology*, 45:205-207, 1953.

[6] S. Ullman. *The Interpretation of Visual Motion*. MIT Press, Cambridge, Mass., 1979.

[7] H.B. Barlow, C. Blakemore, and J.D. Pettigrew. The mechanism of directionally selective units in rabbit's retina. *J. Physiology*, London, 193:327-342, 1967.

[8] H.B. Barlow and W.R. Levick. The mechanism of directionally selective units in rabbit's retina. *J. Physiology*, London, 173:377-407, 1965.

[9] B. Bridgeman. Visual receptive fields to absolute and relative motion during tracking. *Science*, 187:1106-1108, 1972.

[10] O.J. Grusser and U. Grusser. Neuronal mechanisms of visual motion perception. In R. Jung, ed., *Handbook of Sensory Physiology*, Vol. 7, Pt. 3, pp. 332-429. Springer-Verlag, New York, 1973.

[11] D.H. Hubel. *Eye, Brain and Vision*. Scientific American Publishers, New York, 1988.

[12] D.H. Hubel and T.N. Wiesel. Receptive fields and functional architecture of monkey striate cortex. *J. Physiology*, London, 195:215-243, 1968.

[13] K. Nakayama and J.M. Loomis. Optical velocity patterns, velocity sensitive neurons and space perception. *Perception*, 3:63-80, 1974.

[14] T. Poggio and W. Reichardt. Considerations on models of movement detection. *Kybernetic*, 13:223-227, 1973.

[15] T. Poggio, W. Reichardt, and K. Hausen. Figure-ground discrimination by relative movement in the visual system of the fly. *Biological Cybernetics*, 46:1-30, 1983.

[16] R. Sekular and E. Levinson. Mechanisms of motion perception. *Psychologica*, 17:38-49, 1974.

[17] R. Sekular and E. Levinson. The perception of moving targets. *Scientific American*, 236:60-73, 1977.

[18] S.M. Anstis. The perception of apparent movement. In H.C. Longuet-Higgins and N.S. Sutherland, eds., *The Psychology of Vision*, pp. 153-167. Royal Society, London, 1980.

[19] O.J. Braddick. A short range process in apparent motion. *Vision Research*, 14:519-527, 1974.

[20] O.J. Braddick. Low level and high level processes in apparent motion. In H.C. Longuet-Higgins and N.S. Sutherland, eds., *The Psychology of Vision*, pp. 137-149. Royal Society, London, 1980.

[21] M.L. Braunstein. Depth perception in rotation dot patterns: Effects of numerousity and perspective. *J. Experimental Psychology*, 64:415-420, 1962.

[22] E.J. Gibson, J.J. Gibson, O.W. Smith, and H. Flock. Motion parallax as a determinant of perceived depth. *J. Experimental Psychology*, 58:40-51, 1959.

[23] J.J. Gibson. *The Ecological Approach to Visual Perception*. Houghton Mifflin, Boston, 1950.

[24] J.J. Gibson. *The Perception of the Visual World*. Houghton Mifflin, Boston, 1950.

[25] J.J. Gibson. *The Senses Considered as Perceptual Systems*. Houghton Mifflin, Boston, 1966.

[26] J.J. Gibson, P. Olum, and F. Rosenblatt. Parallax and perspective during aircraft landings. *Am. J. Psychology*, 68:372-385, 1955.

[27] G. Johansson. Monocular movement parallax and near-space perception. *Perception*, 2:135-146, 1973.

[28] G. Johansson. Visual perception for biological motion and a model for its analysis. *Perception and Psychophysics*, 14:201-211, 1973.

[29] G. Johansson. Visual motion perception. *Scientific American*, 232(6):76-88, 1975.

[30] L. Kauffman, I. Cyrulnik, J. Kaplowitz, and G. Melnick. The complementarity of apparent and real motion. *Psychol. Forsch.*, 34:343-348, 1971.

[31] D.N. Lee. The optic flow field: The foundation of vision. In H.C. Longuet-Higgins and N.S. Sutherland, eds., *The Psychology of Vision*, pp. 153-167. Royal Society, London, 1980.

[32] D. Regan. Visual processing of four kinds of relative motion. *Vision Research*, 28:127-145, 1986.

[33] D. Regan, K.I. Beverly, and M. Cynader. The visual perception of motion in depth. *Scientific American*, 241:122-133, 1979.

[34] B.K.P Horn and B. Schunck. Determining optical flow. *Artificial Intelligence*, 17:185-203, 1981.

[35] J.J. Koenderink and A.J. van Doorn. Invariant properties of the motion parallax field due to the movement of rigid bodies relative to an observer. *Optica Acta*, 22:773-791, 1975.

[36] J.J. Koenderink and A.J. van Doorn. Local structure of motion parallax of the plane. *J. Optical Soc. America*, 66:717-723, 1976.

[37] J.J. Koenderink and A.J. van Doorn. How an ambulant observer can construct a model of the environment from the geometrical structure of the visual inflow. *Kybernetic*, 77:224-247, 1977.

[38] J.J. Koenderink and A.J. van Doorn. Exterospecific component for the detection of structure and motion in three dimensions. *J. Optical Soc. America*, 71:953-957, 1981.

[39] H.C. Longuet-Higgins and K. Prazdny. The interpretation of a moving retinal image. *Proc. Royal Soc.*, London, B-208:385-397, 1980.

[40] D. Marr and S. Ullman. The role of directional selectivity in early visual processing. *Proc. Royal Soc.*, London, B-211:151-180, 1981.

[41] K. Prazdny. Egomotion and relative depth from optical flow. *Biological Cybernetics*, 36:87-102, 1980.

[42] K. Prazdny. Determining the instantaneous direction of motion from optic flow generated by a curvilinearly moving observer. *Computer Vision, Graphics and Image Processing*, 17:238-248, 1981.

[43] K. Prazdny. A note on perception of surface slant and edge labels from optic flow. *Perception*, 10:579-582, 1981.

[44] K. Prazdny. On the information in optical flows. *Computer Vision, Graphics and Image Processing*, 22:239-259, 1983.

[45] A.M. Waxman. Image flow theory: A framework for 3D inference from time varying imagery. In C. Brown, ed., *Advances in Computer Vision*, pp. 165-223. Lawrence Erlbaum Associates, Publishers, Hillsdale, N.J., 1988.

[46] A. Netravali and J.D. Robbins. Motion compensated TV coding. *BSTJ*, 58:631-670, 1979.

[47] M. Ziegler. Hierarchical motion estimation using the phase correlation method in 140mb/s HDTV coding. In *Proc. Third Int'l Workshop on HDTV*, Torino, Italy, 1989.

[48] H.H. Baker and R.C. Bolles. Generalizing epipolar plane image analysis on the spatiotemporal surface. *Int'l J. Computer Vision*, 3:33-49, 1989.

[49] E.D. Dickmanns. Subject-object discrimination in 4D dynamic scene interpretation of machine vision. In *Proc. IEEE Workshop on Visual Motion*, Irvine, Calif., pp. 298-304, 1989.

[50] K. Skifstad and R. Jain. Range estimation from intensity gradient analysis. *Machine Vision Applications*, 2:81-102, 1989.

[51] D. Marr. *Vision*. W.H. Freeman, New York, 1982.

[52] D. Marr and T. Poggio. A computational theory of human stereo vision. *Proc. Royal Soc.*, London, B-204:301-308, 1979.

[53] S.T. Barnard and W.B. Thompson. Disparity analysis of images. *IEEE Trans. Pattern Analysis and Machine Intelligence*, PAMI-2:333-340, 1980.

[54] L.S. Davis and S. Yam. A generalized Hough-like transformation for shape recognition. Tech. Report TR-134, Dept. of Computer Science, Univ. of Texas, Austin, 1980.

[55] L.S. Dreschler and H.H. Nagel. Volumetric model and 3D trajectory of a moving car derived from a monocular TV-frame sequence of a street scene. *Computer Vision, Graphics and Image Processing*, 20:199-228, 1982.

[56] B. Radig, R. Kraasch, and W. Zack. Matching symbolic descriptions for 3D reconstruction of simple moving objects. In *Proc. IEEE ICPR Conf.*, Miami, Fla., pp. 1081-1084, 1980.

[57] S. Ranade and A. Rosenfeld. Point pattern matching by relaxation. *Pattern Recognition*, 12:269-275, 1980.

[58] A.M. Waxman, J. Wu, and M. Siebert. Computing visual motion in the short and the long: From receptive fields to neural networks. In *Proc. IEEE Workshop on Visual Motion*, Irvine, Calif., 1989.

[59] T.J. Broida and Chellappa. Experiments on uniqueness results on object structure and kinematics from a sequence of monocular images. In *Proc. IEEE Workshop on Visual Motion*, Irvine, Calif., pp. 21-30, 1989.

[60] J.Q. Fang and T.S. Huang. Some experiments on estimating the 3D motion parameters of a rigid body from two consecutive image frames. *IEEE Trans. Pattern Analysis and Machine Intelligence*, PAMI-6:545-554, 1984.

[61] D. Marimont. *Inferring Spatial Structure from Feature Correspondences*. PhD thesis, Dept. of Electrical Engineering, Stanford Univ., Stanford, Calif., 1986.

[62] J.W. Roach and J.K. Aggarwal. Determining the movement of objects from a sequence of images. *IEEE Trans. Pattern Analysis and Machine Intelligence*, PAMI-2:554-562, 1980.

[63] R.Y. Tsai. Multiframe image point matching and 3D surface reconstruction. *IEEE Trans. Pattern Analysis and Machine Intelligence*, PAMI-5:159-174, 1983.

[64] R.Y. Tsai and T.S. Huang. Uniqueness and estimation of three-dimensional motion parameters of rigid objects with curved surfaces. *IEEE Trans. Pattern Analysis and Machine Intelligence*, PAMI-6:13-27, 1984.

[65] J. Weng, T.S. Huang, and N. Ahuja. Motion and structure from two perspective views: Algorithms and error analysis. *IEEE Trans. Pattern Analysis and Machine Intelligence*, PAMI-11:451-476, 1989.

[66] B.F. Buxton and H. Buxton. Computation of optic flow from the motion of edge features in image sequences. *Image and Vision Computing*, 2:59-75, 1984.

[67] N. Cornelius and T. Kanade. Adapting optical flow to measure object motion in reflectance and x-ray image sequences. In *Proc. ACM Siggraph/Sigart Interdisciplinary Workshop on Motion*, Toronto, pp. 50-58, 1983.

[68] W. Enkelmann. Investigations of multigrid algorithms for estimation of optical flow fields in image sequences. *Computer Vision, Graphics and Image Processing*, 43:150-177, 1988.

[69] C.L. Fennema and W.B. Thompson. Velocity determination in scenes containing several moving objects. *Computer Vision, Graphics and Image Processing*, 9:301-315, 1979.

[70] F. Glazer. *Hierarchical Motion Detection*. PhD thesis, COINS Dept., Univ. of Massachusetts, Amherst, 1987.

[71] E.C. Hildreth. *The Measurement of Visual Motion*. MIT Press, Cambridge, Mass., 1983.

[72] J.O. Limb and J.A. Murphy. Estimating velocities of moving images in television signals. *Computer Graphics and Image Processing*, 4:311-327, 1975.

[73] H.H. Nagel. Displacement vectors derived from second order intensity variations in image sequences. *Computer Vision, Graphics and Image Processing*, 21:85-117, 1983.

[74] H.H. Nagel. On the estimation of dense displacement maps from image sequences. In *Proc. ACM Motion Workshop*, Toronto, pp. 59-65, 1983.

[75] H.H. Nagel and W. Enkelmann. An estimation of smoothness constraints for the estimation of displacement vectors fields from image sequences. *IEEE Trans. Pattern Analysis and Machine Intelligence*, PAMI-8:565-593, 1986.

[76] B. Schunck. Image flow: Fundamentals and algorithms. In J.K. Martin and W.N. Aggarwal, eds., *Motion Understanding: Robot and Human Vision*, pp. 23-68. Kluwer Academic Publishers, Boston, 1988.

[77] W.B. Thompson and S.T. Barnard. Lower level estimation and interpretation of visual motion. *Computer*, 20(8):20-28, 1987.

[78] A.M. Waxman, J. Wu, and F. Bergholm. Convected activation profiles and measurement of visual motion. In *Proc. IEEE CVPR*, Ann Arbor, Mich., pp. 717-722, 1988.

[79] M. Yachida. Determining velocity maps by spatiotemporal neighborhoods from image sequences. *Computer Vision, Graphics and Image Processing*, 21:262-279, 1983.

[80] P. Anandan. *Measuring Visual Motion from Image Sequences*. PhD thesis, COINS Dept., Univ. of Massachusetts, Amherst, 1987.

[81] P.J. Burt, C. Yen, and X. Xu. Multi-resolution flow-through motion analysis. In *Proc. IEEE CVPR Conf.*, pp. 246-252, 1983.

[82] F. Glazer, G. Reynolds, and P. Anandan. Scene matching by hierarchical correlation. In *Proc. IEEE CVPR*, pp. 432-441, 1983.

[83] G.L. Scott. Four-line method of locally estimating optic flow. *Image and Vision Computing*, 5(2), 1986.

[84] R.Y Wong and E.L. Hall. Sequential hierarchical scene matching. *IEEE Trans. Computers*, 27(4):359-366, 1978.

[85] E.H. Adelson and J.R. Bergen. Spatiotemporal energy models for the perception of motion. *J. Optical Soc. America*, 2:284-299, 1985.

[86] E.H. Adelson and J.R. Bergen. The extraction of spatiotemporal energy in human and machine vision. In *Proc. IEEE Workshop on Motion*, Charleston, S.C., pp. 151-155, 1986.

[87] D.J. Fleet and A.D. Jepson. On hierarchical construction of orientation and velocity selective filters. Tech. Report RBCV-TR-85-8, Dept. of Computer Science, Univ. of Toronto, 1985.

[88] D. Heeger. A model for extraction of image flow. In *Proc. First Int'l Conf. Computer Vision*, London, pp. 181-190, 1987.

[89] J.P.H. Van Santen and G. Sperling. Elaborated Reichardt detectors. *J. Optical Soc. America*, 2(2):300-321, 1985.

[90] A.B. Watson and A.J. Ahumada. A look at motion in frequency domain. NASA Tech. Memorandum 84352, NASA Ames Research Center, Moffett Field, Calif., 1983.

[91] A.B. Watson and A.J. Ahumada. Model of human visual motion sensing. *J. Optical Soc. America*, 2:322-341, 1985.

[92] S.P. Liou and R. Jain. Motion detection in spatiotemporal space. *Computer Vision, Graphics and Image Processing*, 45:227-250, 1989.

[93] W.F. Clocksin. Perception of surface slant and edge labels from optic flow. *Perception*, 9:253-269, 1980.

[94] D.N. Lee. Visual information during locomotion. In I.D.G. MacLed and O. Pick, eds., *Perception: Essays in Honor of James J. Gibson*. Cornell Univ. Press, Cornell Univ., Ithaca, N.Y., 1974.

[95] D.W. Murray, D.A. Castelow, and B.F. Buxton. From image sequences to recognized moving polyhedral objects. *Int'l J. Computer Vision* 3(3), 1989.

[96] R.C. Nelson and J. Aloimonos. Obstacle avoidance using flow-field divergence. *IEEE Trans. Pattern Analysis and Machine Intelligence*, PAMI-11:1102-1106, 1989.

[97] B. Shahraray and M.K. Brown. Robust depth estimation from optic flow. In *Proc. IEEE CVPR Conf.*, Ann Arbor, Mich., pp. 641-649, 1988.

[98] M. Subbarao. *Interpretation of Visual Motion: A Computational Study*. PhD thesis, Center for Automation Research, Univ. of Maryland, College Park, 1988.

[99] M. Subbarao and A.M. Waxman. On the uniqueness of image-flow solutions for planar surfaces in motion. In *Proc. IEEE Workshop on Computer Vision*, Bellaire, Mich., pp. 129-140, 1985.

[100] A.M. Waxman, B. Kamgar-Parsi, and M. Subbrao. Closed form solutions to image-flow equations. *Int'l J. Computer Vision*, 1:239-258, 1987.

[101] A.M. Waxman and S. Ullman. Surface structure and three dimensional motion from image-flow kinematics. *Int'l J. Robotics*, 4:72-94, 1985.

[102] A.M. Waxman and K. Wohn. Contour evolution, neighborhood deformation and global image flow: Planar surfaces in motion. *Int'l J. Robotics*, 4:95-108, 1985.

[103] J.K. Aggarwal and N. Nandhakumar. On the computation of motion from sequences of images—a review. Tech. Report TR-88-2-47, Computer Vision Research Center, Univ. of Texas, Austin, 1988.

[104] J. Barron. A survey of approaches for determining optic flow, environmental layout and egomotion. Tech. Report RBCV-TR-84-5, Computer Science Dept., Univ. of Toronto, 1984.

[105] D. Ballard and C. Brown. *Computer Vision*. Prentice-Hall, Englewood Cliffs, N.J., 1982.

[106] A. Verri and T. Poggio. Against quantitative optical flow. In *Proc. First Int'l Conf. Computer Vision*, London, pp. 171-180, 1987.

[107] M. Born and E. Wolf. *Principles of Optics*. Pergamon Press, New York, 1959.

[108] B.K.P. Horn and R.W. Sjoberg. Calculating the reflectance map. *Applied Optics*, 18:1170-1779, 1979.

[109] M.P. deCarmo. *Differential Geometry of Curves and Surfaces*. Prentice-Hall, Englewood Cliffs, N.J., 1976.

[110] D.W. Murray and B.F. Buxton. *Experiments in the Machine Interpretation of Visual Motion*. MIT Press, Cambridge, Mass., 1990.

[111] A.R. Bruss and B.K.P. Horn. Passive navigation. *Computer Vision, Graphics and Image Processing*, 21:3-20, 1983.

[112] D. Heeger and A. Jepson. Simple method for computing 3D motion and depth. In *Proc. Third Int'l Conf. Computer Vision*, Osaka, Japan, pp. 96-100, 1990.

[113] G. Adiv. Inherent ambiguities in recovering 3D motion and structure from a noisy flow field. In *Proc. IEEE Computer Vision and Pattern Recognition Conf.*, pp. 70-77, 1985.

[114] T.C. Chou and K. Kanatani. Recovering 3D rigid motions without correspondence. In *Proc. First Int'l Conf. Computer Vision*, London, pp. 534-538, 1987.

[115] B.K.P. Horn and E.J. Weldon. Computationally efficient methods for recovering translational motion. In *Proc. First Int'l Conf. Computer Vision*, London, pp. 2-11, 1987.

[116] D.T. Lawton. Processing translational motion sequences. *Computer Vision, Graphics and Image Processing*, 22:116-144, 1983.

[117] L. Matthies, R. Szeliski, and T. Kanade. Kalman filter-based algorithms for estimating depth from image-sequences. In *Proc. Second Int'l Conf. Computer Vision*, Tampa, Fla., pp. 199-213, 1988.

[118] S. Negahdaripour and B.K.P. Horn. Direct passive navigation. *IEEE Trans. Pattern Analysis and Machine Intelligence*, PAMI-9:512-522, 1987.

[119] J.H. Rieger and D.T. Lawton. Determining the instantaneous axis of translation from optic flow generated by arbitrary sensor motion. In *Proc. ACM Interdisciplinary Workshop on Motion,* Toronto, pp. 33-41, 1983.

[120] C. Tomasi and T. Kanade. Shape and motion without depth. In *Proc. Third Int'l Conf. Computer Vision,* Osaka, Japan, pp. 90-95, 1990.

[121] T.D. Williams. Depth from camera motion in a real-world scene. *IEEE Trans. Pattern Analysis and Machine Intelligence,* PAMI-2:511-516, 1980.

[122] B.I. Justusson. Median filtering: Statistical properties. In T.S. Huang, ed., *Two Dimensional Digital Signal Processing*, pp. 161-196. Springer-Verlag, Berlin, 1981.

[123] T.S. Huang and Y.P. Hsu. Image sequence enhancement. In T.S. Huang, ed., *Image Sequence Analysis*, pp. 290-310. Springer-Verlag, Berlin, 1981.

[124] D.S. Kalivas and A.A. Sawchuck. Segmentation, motion estimation and enhancement of noisy image-sequences. *IEEE Trans. Pattern Analysis and Machine Intelligence* (submitted), 1991.

[125] G. Wolberg. *Digital Image Warping*. IEEE Computer Society Press, Los Alamitos, Calif., 1990.

[126] W.E.L. Grimson. *From Images to Surfaces: A Computational Study of the Human Early Visual System.* MIT Press, Cambridge, Mass., 1981.

[127] B.D. Lucas and T. Kanade. An iterative image registration technique with an application to stereo vision. In *Proc. IJCAI*, pp. 674-679, 1981.

[128] H.P. Moravec. Towards automatic visual obstacle avoidance. In *Proc. IJCAI*, pp. 584-585, 1977.

[129] K. Wohn, L.S. Davis, and P. Thrift. Motion estimation based on multiple local constraints and non-linear smoothing. *Pattern Recognition,* 16:563-570, 1983.

[130] R. Goldenberg, W.C. Lau, A. She, and A.M. Waxman. Progress on the prototype pipe. In *Proc. IEEE Robotics and Automation Conf.*, 1987.

[131] E.W. Kent, M. Shneier, and R. Lumia. Pipelined image processing engine. *J. Parallel and Distributed Computing*, 2:50-78, 1985.

[132] A. Singh. Image processing on the PIPE. Tech. Report TN-87-093, Philips Laboratories, Briarcliff Manor, N.Y., 1987.

[133] D. Gennery. *Modeling the Environment of an Exploring Vehicle by Means of Stereo Vision.* PhD thesis, Stanford Artificial Intelligence Laboratory, Stanford Univ., Stanford, Calif., 1980.

[134] L.H. Quam. Hierarchical warp stereo. In *Proc. DARPA Image Understanding Workshop*, pp. 149-156, 1984.

[135] G.L. Scott. *Local and Global Interpretation of Moving Images*. Morgan Kaufmann, 1988.

[136] P.J. Burt. The pyramid as a structure for efficient computation. In A. Rosenfeld, ed., *Multi Resolution Image Processing and Analysis*, pp. 6-37. Springer-Verlag, Berlin, 1984.

[137] J.K. Tsotsos, D.J. Fleet, and A.D. Jepson. Towards a theory for motion understanding in man and machine. In W.N. Martin and J.K. Aggarwal, eds., *Motion Understanding: Robot and Human Vision*, pp. 353-418. Kluwer Academic Publishers, Boston, 1988.

[138] D. Marr and E.C. Hildreth. Theory of edge detection. *Proc. Royal Soc.*, London, B-207:187-217, 1980.

[139] J.K. Kearney and W.B. Thompson. Bounding constraint propagation for optical-flow estimation. In W.N. Martin and J.K. Aggarwal, eds., *Motion Understanding: Robot and Human Vision*, pp. 1-22. Kluwer Academic Publishers, Boston, 1988.

[140] J. Aisbett. Optical-flow with an intensity weighted smoothing. *IEEE Trans. Pattern Analysis and Machine Intelligence*, PAMI-5:512-522, 1989.

[141] J. Hutchinson, K. Koch, and C. Mead. Computing motion using analog and binary resistive networks. *Computer*, 21(3):52-63, 1988.

[142] D. Shulman and J. Herve. Regularization of discontinuous flow fields. In *Proc. IEEE-ACM Workshop on Visual Motion*, pp. 81-86, 1989.

[143] D.W. Murray and B.F. Buxton. Reconstructing the optic flow field from edge motion: An examination of two different approaches. In *First Conf. AI Applications*, Denver, 1984.

[144] J.V. Beck and K.J. Arnold. *Parameter estimation in engineering and science*. John Wiley, New York, 1977.

[145] A. Singh. An estimation-theoretic framework for image flow computation. In *Proc. Third Int'l Conf. Computer Vision*, Osaka, Japan, pp. 168-177, 1990.

[146] G.B. Thomas. *Calculus and Analytical Geometry*. Addison-Wesley, Reading, Mass., 1968.

[147] A. Ralston and P. Rabinowitz. *A First Course in Numerical Analysis*. McGraw-Hill, New York, 1978.

[148] M.A. Snyder. On the mathematical foundations of smoothness constraints for the determination of optical flow and for surface reconstruction. In *Proc. IEEE Workshop on Visual Motion*, 1989, pp. 107-115, 1989.

[149] G.H. Golub and C.F. Van Loan. *Matrix Computations*. Johns Hopkins Univ. Press, Baltimore, Md., 1983.

[150] W.H. Press et al. *Numerical Recipes in C*. Cambridge Univ. Press, Cambridge, U.K., 1988.

[151] S. Geman and D. Geman. Stochastic relaxation, Gibbs distribution and the Bayesian restoration of images. *IEEE Trans. Pattern Analysis and Machine Intelligence*, 6:721-741, 1984.

[152] R. Szeliski. *Bayesian Model of Uncertainty in Low-Level Vision*. PhD thesis, Dept. of Computer Science, Carnegie Mellon Univ., Pittsburgh, 1988.

[153] J.J. Tuma. *Engineering Mathematics Handbook*. McGraw-Hill, New York, 1979.

[154] P. Beaudet. Rotationally invariant image operators. In *Proc. Int'l Conf. Pattern Recognition*, pp. 579-583, 1978.

[155] D. Gabor. Theory of communication. *J. IEE*, London, 93:429-457, 1946.

[156] S.G. Mallat. A theory for multi-resolution signal decomposition: The wavelet representation. Tech. Report MS-CIS-87-22, Dept. of Computer and Information Science, Univ. of Pennsylvania, Philadelphia, 1987.

[157] A. Singh and P.K. Allen. A hierarchical model for optic flow extraction via spatiotemporal frequency channels. In *Proc. DARPA Image Understanding Workshop*, Boston, pp. 961-969, 1988.

[158] A. Gelb, ed., *Applied optimal estimation.* MIT Press, Cambridge, Mass., 1988.

[159] A. Singh. A method to integrate correlation-based and gradient-based methods for image-flow estimation. In *Proc. SPIE Conf. Advances in Intelligent Robotics Systems*, Philadelphia, 1989.

Bibliography

E.H. Adelson and J.R. Bergen. Spatiotemporal energy models for the perception of motion. *J. Optical Soc. America*, 2:284-299, 1985.

E.H. Adelson and J.R. Bergen. The extraction of spatiotemporal energy in human and machine vision. In *Proc. IEEE Workshop on Motion*, Charleston, S.C., pp. 151-155, 1986.

G. Adiv. Inherent ambiguities in recovering 3D motion and structure from a noisy flow field. In *Proc. IEEE Computer Vision and Pattern Recognition Conf.*, pp. 70-77, 1985.

J.K. Aggarwal and N. Nandhakumar. On the computation of motion from sequences of images—a review. Tech. Report TR-88-2-47, Computer Vision Research Center, Univ. of Texas, Austin, 1988.

J. Aisbett. Optical-flow with an intensity weighted smoothing. *IEEE Trans. Pattern Analysis and Machine Intelligence*, PAMI-5:512-522, 1989.

P. Anandan. *Measuring Visual Motion from Image Sequences.* PhD thesis, COINS Dept., Univ. of Massachusetts, Amherst, 1987.

S.M. Anstis. The perception of apparent movement. In H.C. Longuet-Higgins and N.S. Sutherland, eds., *The Psychology of Vision*, pp. 153-167. Royal Society, London, 1980.

H.H. Baker and R.C. Bolles. Generalizing epipolar plane image analysis on the spatiotemporal surface. *Int'l J. Computer Vision*, 3:33-49, 1989.

D. Ballard and C. Brown. *Computer Vision.* Prentice-Hall, Englewood Cliffs, N.J., 1982.

H.B. Barlow, C. Blakemore, and J.D. Pettigrew. The mechanism of directionally selective units in rabbit's retina. *J. Physiology*, London, 193:327-342, 1967.

H.B. Barlow and W.R. Levick. The mechanism of directionally selective units in rabbit's retina. *J. Physiology*, London, 173:377-407, 1965.

S.T. Barnard and W.B. Thompson. Disparity analysis of images. *IEEE Trans. Pattern Analysis and Machine Intelligence*, PAMI-2:333-340, 1980.

J. Barron. A survey of approaches for determining optic flow, environmental layout and egomotion. Tech. Report RBCV-TR-84-5, Computer Science Dept., Univ. of Toronto, 1984.

P. Beaudet. Rotationally invariant image operators. In *Proc. Int'l Conf. Pattern Recognition*, pp. 579-583, 1978.

J.V. Beck and K.J. Arnold. *Parameter estimation in engineering and science.* John Wiley, New York, 1977.

M. Born and E. Wolf. *Principles of Optics.* Pergamon Press, New York, 1959.

O.J. Braddick. A short range process in apparent motion. *Vision Research*, 14:519-527, 1974.

O.J. Braddick. Low level and high level processes in apparent motion. In H.C. Longuet-Higgins and N.S. Sutherland, eds., *The Psychology of Vision*, pp. 137-149. Royal Society, London, 1980.

M.L. Braunstein. Depth perception in rotation dot patterns: Effects of numerousity and perspective. *J. Experimental Psychology*, 64:415-420, 1962.

B. Bridgeman. Visual receptive fields to absolute and relative motion during tracking. *Science*, 187:1106-1108, 1972.

T.J. Broida and Chellappa. Experiments on uniqueness results on object structure and kinematics from a sequence of monocular images. In *Proc. IEEE Workshop on Visual Motion*, Irvine, Calif., pp. 21-30, 1989.

A.R. Bruss and B.K.P. Horn. Passive navigation. *Computer Vision, Graphics and Image Processing*, 21:3-20, 1983.

P.J. Burt. The pyramid as a structure for efficient computation. In A. Rosenfeld, ed., *Multi Resolution Image Processing and Analysis*, pp. 6-37. Springer-Verlag, Berlin, 1984.

P.J. Burt, C. Yen, and X. Xu. Multi-resolution flow-through motion analysis. In *Proc. IEEE CVPR Conf.*, pp. 246-252, 1983.

H.E. Burton. The optics of Euclid. *J. Optical Soc. America*, 35:357-372, 1945.

B.F. Buxton and H. Buxton. Computation of optic flow from the motion of edge features in image sequences. *Image and Vision Computing*, 2:59-75, 1984.

T.C. Chou and K. Kanatani. Recovering 3D rigid motions without correspondence. In *Proc. First Int'l Conf. Computer Vision*, London, pp. 534-538, 1987.

W.F. Clocksin. Perception of surface slant and edge labels from optic flow. *Perception*, 9:253-269, 1980.

N. Cornelius and T. Kanade. Adapting optical flow to measure object motion in reflectance and x-ray image sequences. In *Proc. ACM Siggraph/Sigart Interdisciplinary Workshop on Motion*, Toronto, pp. 50-58, 1983.

J.E. Cutting. Four assumptions about invariance in perception. *J. Experimental Psychology: Human Perception and Performance*, 9:310-317, 1983.

L.S. Davis and S. Yam. A generalized Hough-like transformation for shape recognition. Tech. Report TR-134, Dept. of Computer Science, Univ. of Texas, Austin, 1980.

M.P. deCarmo. *Differential Geometry of Curves and Surfaces.* Prentice-Hall, Englewood Cliffs, N.J., 1976.

E.D. Dickmanns. Subject-object discrimination in 4D dynamic scene interpretation of machine vision. In *Proc. IEEE Workshop on Visual Motion*, Irvine, Calif., pp. 298-304, 1989.

L.S. Dreschler and H.H. Nagel. Volumetric model and 3D trajectory of a moving car derived from a monocular TV-frame sequence of a street scene. *Computer Vision, Graphics and Image Processing*, 20:199-228, 1982.

W. Enkelmann. Investigations of multigrid algorithms for estimation of optical flow fields in image sequences. *Computer Vision, Graphics and Image Processing*, 43:150-177, 1988.

J.Q. Fang and T.S. Huang. Some experiments on estimating the 3D motion parameters of a rigid body from two consecutive image frames. *IEEE Trans. Pattern Analysis and Machine Intelligence*, PAMI-6:545-554, 1984.

C.L. Fennema and W.B. Thompson. Velocity determination in scenes containing several moving objects. *Computer Vision, Graphics and Image Processing*, 9:301-315, 1979.

D.J. Fleet and A.D. Jepson. On hierarchical construction of orientation and velocity selective filters. Tech. Report RBCV-TR-85-8, Dept. of Computer Science, Univ. of Toronto, 1985.

D. Gabor. Theory of communication. *J. IEE*, London, 93:429-457, 1946.

A. Gelb, ed., *Applied optimal estimation.* MIT Press, Cambridge, Mass., 1988.

S. Geman and D. Geman. Stochastic relaxation, Gibbs distribution and the Bayesian restoration of images. *IEEE Trans. Pattern Analysis and Machine Intelligence*, 6:721-741, 1984.

D. Gennery. *Modeling the Environment of an Exploring Vehicle by Means of Stereo Vision.* PhD thesis, Stanford Artificial Intelligence Laboratory, Stanford Univ., Stanford, Calif., 1980.

E.J. Gibson, J.J. Gibson, O.W. Smith, and H. Flock. Motion parallax as a determinant of perceived depth. *J. Experimental Psychology*, 58:40-51, 1959.

J.J. Gibson. *The Ecological Approach to Visual Perception.* Houghton Mifflin, Boston, 1950.

J.J. Gibson. *The Perception of the Visual World.* Houghton Mifflin, Boston, 1950.

J.J. Gibson. *The Senses Considered as Perceptual Systems.* Houghton Mifflin, Boston, 1966.

J.J. Gibson, P. Olum, and F. Rosenblatt. Parallax and perspective during aircraft landings. *Am. J. Psychology*, 68:372-385, 1955.

F. Glazer. *Hierarchical Motion Detection.* PhD thesis, COINS Dept., Univ. of Massachusetts, Amherst, 1987.

F. Glazer, G. Reynolds, and P. Anandan. Scene matching by hierarchical correlation. In *Proc. IEEE CVPR*, pp. 432-441, 1983.

R. Goldenberg, W.C. Lau, A. She, and A.M. Waxman. Progress on the prototype pipe. In *Proc. IEEE Robotics and Automation Conf.*, 1987.

G.H. Golub and C.F. Van Loan. *Matrix Computations.* Johns Hopkins Univ. Press, Baltimore, Md., 1983.

W.E.L. Grimson. *From Images to Surfaces: A Computational Study of the Human Early Visual System.* MIT Press, Cambridge, Mass., 1981.

O.J. Grusser and U. Grusser. Neuronal mechanisms of visual motion perception. In R. Jung, ed., *Handbook of Sensory Physiology,* Vol. 7, Pt. 3, pp. 332-429. Springer-Verlag, New York, 1973.

D. Heeger. A model for extraction of image flow. In *Proc. First Int'l Conf. Computer Vision*, London, pp. 181-190, 1987.

D. Heeger and A. Jepson. Simple method for computing 3D motion and depth. In *Proc. Third Int'l Conf. Computer Vision*, Osaka, Japan, pp. 96-100, 1990.

H. von Helmholtz. *Physiological Optics*, Vol. 3, third ed., J.P.C. Southall, trans., Optical Soc. America, Menasha, Wis., 1925.

E.C. Hildreth. *The Measurement of Visual Motion.* MIT Press, Cambridge, Mass., 1983.

B.K.P Horn and B. Schunck. Determining optical flow. *Artificial Intelligence*, 17:185-203, 1981.

B.K.P. Horn and R.W. Sjoberg. Calculating the reflectance map. *Applied Optics*, 18:1170-1779, 1979.

B.K.P. Horn and E.J. Weldon. Computationally efficient methods for recovering translational motion. In *Proc. First Int'l Conf. Computer Vision*, London, pp. 2-11, 1987.

T.S. Huang and Y.P. Hsu. Image sequence enhancement. In T.S. Huang, ed., *Image Sequence Analysis*, pp. 290-310. Springer-Verlag, Berlin, 1981.

D.H. Hubel. *Eye, Brain and Vision.* Scientific American Publishers, New York, 1988.

D.H. Hubel and T.N. Wiesel. Receptive fields and functional architecture of monkey striate cortex. *J. Physiology*, London, 195:215-243, 1968.

J. Hutchinson, K. Koch, and C. Mead. Computing motion using analog and binary resistive networks. *Computer*, 21(3):52-63, 1988.

G. Johansson. Monocular movement parallax and near-space perception. *Perception*, 2:135-146, 1973.

G. Johansson. Visual perception for biological motion and a model for its analysis. *Perception and Psychophysics*, 14:201-211, 1973.

G. Johansson. Visual motion perception. *Scientific American*, 232(6):76-88, 1975.

B.I. Justusson. Median filtering: Statistical properties. In T.S. Huang, ed., *Two Dimensional Digital Signal Processing*, pp. 161-196. Springer-Verlag, Berlin, 1981.

D.S. Kalivas and A.A. Sawchuck. Segmentation, motion estimation and enhancement of noisy image-sequences. *IEEE Trans. Pattern Analysis and Machine Intelligence* (submitted), 1991.

L. Kauffman, I. Cyrulnik, J. Kaplowitz, and G. Melnick. The complementarity of apparent and real motion. *Psychol. Forsch.*, 34:343-348, 1971.

J.K. Kearney and W.B. Thompson. Bounding constraint propagation for optical-flow estimation. In W.N. Martin and J.K. Aggarwal, eds., *Motion Understanding: Robot and Human Vision*, pp. 1-22. Kluwer Academic Publishers, Boston, 1988.

E.W. Kent, M. Shneier, and R. Lumia. Pipelined image processing engine. *J. Parallel and Distributed Computing*, 2:50-78, 1985.

J.J. Koenderink and A.J. van Doorn. Invariant properties of the motion parallax field due to the movement of rigid bodies relative to an observer. *Optica Acta*, 22:773-791, 1975.

J.J. Koenderink and A.J. van Doorn. Local structure of motion parallax of the plane. *J. Optical Soc. America*, 66:717-723, 1976.

J.J. Koenderink and A.J. van Doorn. How an ambulant observer can construct a model of the environment from the geometrical structure of the visual inflow. *Kybernetic*, 77:224-247, 1977.

J.J. Koenderink and A.J. van Doorn. Exterospecific component for the detection of structure and motion in three dimensions. *J. Optical Soc. America*, 71:953-957, 1981.

D.T. Lawton. Processing translational motion sequences. *Computer Vision, Graphics and Image Processing*, 22:116-144, 1983.

D.N. Lee. Visual information during locomotion. In I.D.G. MacLed and O. Pick, eds., *Perception: Essays in Honor of James J. Gibson.* Cornell Univ. Press, Cornell Univ., Ithaca, N.Y., 1974.

D.N. Lee. The optic flow field: The foundation of vision. In H.C. Longuet-Higgins and N.S. Sutherland, eds., *The Psychology of Vision*, pp. 153-167. Royal Society, London, 1980.

J.O. Limb and J.A. Murphy. Estimating velocities of moving images in television signals. *Computer Graphics and Image Processing*, 4:311-327, 1975.

S.P. Liou and R. Jain. Motion detection in spatiotemporal space. *Computer Vision, Graphics and Image Processing*, 45:227-250, 1989.

H.C. Longuet-Higgins and K. Prazdny. The interpretation of a moving retinal image. *Proc. Royal Soc.*, London, B-208:385-397, 1980.

B.D. Lucas and T. Kanade. An iterative image registration technique with an application to stereo vision. In *Proc. IJCAI*, pp. 674-679, 1981.

S.G. Mallat. A theory for multi-resolution signal decomposition: The wavelet representation. Tech. Report MS-CIS-87-22, Dept. of Computer and Information Science, Univ. of Pennsylvania, Philadelphia, 1987.

D. Marimont. *Inferring Spatial Structure from Feature Correspondences.* PhD thesis, Dept. of Electrical Engineering, Stanford Univ., Stanford, Calif., 1986.

D. Marr. *Vision.* W.H. Freeman, New York, 1982.

D. Marr and E.C. Hildreth. Theory of edge detection. *Proc. Royal Soc.*, London, B-207:187-217, 1980.

D. Marr and T. Poggio. A computational theory of human stereo vision. *Proc. Royal Soc.*, London, B-204:301-308, 1979.

D. Marr and S. Ullman. The role of directional selectivity in early visual processing. *Proc. Royal Soc.*, London, B-211:151-180, 1981.

L. Matthies, R. Szeliski, and T. Kanade. Kalman filter-based algorithms for estimating depth from image-sequences. In *Proc. Second Int'l Conf. Computer Vision,* Tampa, Fla., pp. 199-213, 1988.

W.R. Miles. Movement in interpretations of the silhouette of a revolving fan. *Am. J. Psychology*, 43:392-404, 1931.

H.P. Moravec. Towards automatic visual obstacle avoidance. In *Proc. IJCAI*, pp. 584-585, 1977.

D.W. Murray and B.F. Buxton. *Experiments in the Machine Interpretation of Visual Motion.* MIT Press, Cambridge, Mass., 1990.

D.W. Murray and B.F. Buxton. Reconstructing the optic flow field from edge motion: An examination of two different approaches. In *First Conf. AI Applications*, Denver, 1984.

D.W. Murray, D.A. Castelow, and B.F. Buxton. From image sequences to recognized moving polyhedral objects. *Int'l J. Computer Vision* 3(3), 1989.

H.H. Nagel. Displacement vectors derived from second order intensity variations in image sequences. *Computer Vision, Graphics and Image Processing*, 21:85-117, 1983.

H.H. Nagel. On the estimation of dense displacement maps from image sequences. In *Proc. ACM Motion Workshop*, Toronto, pp. 59-65, 1983.

H.H. Nagel and W. Enkelmann. An estimation of smoothness constraints for the estimation of displacement vectors fields from image sequences. *IEEE Trans. Pattern Analysis and Machine Intelligence*, PAMI-8:565-593, 1986.

K. Nakayama and J.M. Loomis. Optical velocity patterns, velocity sensitive neurons and space perception. *Perception*, 3:63-80, 1974.

S. Negahdaripour and B.K.P. Horn. Direct passive navigation. *IEEE Trans. Pattern Analysis and Machine Intelligence*, PAMI-9:512-522, 1987.

R.C. Nelson and J. Aloimonos. Obstacle avoidance using flow-field divergence. *IEEE Trans. Pattern Analysis and Machine Intelligence*, PAMI-11:1102-1106, 1989.

A. Netravali and J.D. Robbins. Motion compensated TV coding. *BSTJ*, 58:631-670, 1979.

T. Poggio and W. Reichardt. Considerations on models of movement detection. *Kybernetic*, 13:223-227, 1973.

T. Poggio, W. Reichardt, and K. Hausen. Figure-ground discrimination by relative movement in the visual system of the fly. *Biological Cybernetics*, 46:1-30, 1983.

K. Prazdny. Egomotion and relative depth from optical flow. *Biological Cybernetics*, 36:87-102, 1980.

K. Prazdny. Determining the instantaneous direction of motion from optic flow generated by a curvilinearly moving observer. *Computer Vision, Graphics and Image Processing*, 17:238-248, 1981.

K. Prazdny. A note on perception of surface slant and edge labels from optic flow. *Perception*, 10:579-582, 1981.

K. Prazdny. On the information in optical flows. *Computer Vision, Graphics and Image Processing*, 22:239-259, 1983.

W.H. Press et al. *Numerical Recipes in C*. Cambridge Univ. Press, Cambridge, U.K., 1988.

L.H. Quam. Hierarchical warp stereo. In *Proc. DARPA Image Understanding Workshop*, pp. 149-156, 1984.

B. Radig, R. Kraasch, and W. Zack. Matching symbolic descriptions for 3D reconstruction of simple moving objects. In *Proc. IEEE ICPR Conf.*, Miami, Fla., pp. 1081-1084, 1980.

A. Ralston and P. Rabinowitz. *A First Course in Numerical Analysis*. McGraw-Hill, New York, 1978.

S. Ranade and A. Rosenfeld. Point pattern matching by relaxation. *Pattern Recognition*, 12:269-275, 1980.

D. Regan. Visual processing of four kinds of relative motion. *Vision Research*, 28:127-145, 1986.

D. Regan, K.I. Beverly, and M. Cynader. The visual perception of motion in depth. *Scientific American*, 241:122-133, 1979.

J.H. Rieger and D.T. Lawton. Determining the instantaneous axis of translation from optic flow generated by arbitrary sensor motion. In *Proc. ACM Interdisciplinary Workshop on Motion*, Toronto, pp. 33-41, 1983.

J.W. Roach and J.K. Aggarwal. Determining the movement of objects from a sequence of images. *IEEE Trans. Pattern Analysis and Machine Intelligence*, PAMI-2:554-562, 1980.

B. Schunck. Image flow: Fundamentals and algorithms. In J.K. Martin and W.N. Aggarwal, eds., *Motion Understanding: Robot and Human Vision*, pp. 23-68. Kluwer Academic Publishers, Boston, 1988.

G.L. Scott. Four-line method of locally estimating optic flow. *Image and Vision Computing*, 5(2), 1986.

G.L. Scott. *Local and Global Interpretation of Moving Images*. Morgan Kaufmann, 1988.

R. Sekular and E. Levinson. Mechanisms of motion perception. *Psychologica*, 17:38-49, 1974.

R. Sekular and E. Levinson. The perception of moving targets. *Scientific American*, 236:60-73, 1977.

B. Shahraray and M.K. Brown. Robust depth estimation from optic flow. In *Proc. IEEE CVPR Conf.*, Ann Arbor, Mich., pp. 641-649, 1988.

D. Shulman and J. Herve. Regularization of discontinuous flow fields. In *Proc. IEEE-ACM Workshop on Visual Motion*, pp. 81-86, 1989.

A. Singh. Image processing on the PIPE. Tech. Report TN-87-093, Philips Laboratories, Briarcliff Manor, N.Y., 1987.

A. Singh. An estimation-theoretic framework for image flow computation. In *Proc. Third Int'l Conf. Computer Vision*, Osaka, Japan, pp. 168-177, 1990.

A. Singh. A method to integrate correlation-based and gradient-based methods for image-flow estimation. In *Proc. SPIE Conf. Advances in Intelligent Robotics Systems*, Philadelphia, 1989.

A. Singh and P.K. Allen. A hierarchical model for optic flow extraction via spatiotemporal frequency channels. In *Proc. DARPA Image Understanding Workshop*, Boston, pp. 961-969, 1988.

K. Skifstad and R. Jain. Range estimation from intensity gradient analysis. *Machine Vision Applications*, 2:81-102, 1989.

M.A. Snyder. On the mathematical foundations of smoothness constraints for the determination of optical flow and for surface reconstruction. In *Proc. IEEE Workshop on Visual Motion*, 1989, pp. 107-115, 1989.

M. Subbarao. *Interpretation of Visual Motion: A Computational Study*. PhD thesis, Center for Automation Research, Univ. of Maryland, College Park, 1988.

M. Subbarao and A.M. Waxman. On the uniqueness of image-flow solutions for planar surfaces in motion. In *Proc. IEEE Workshop on Computer Vision*, Bellaire, Mich., pp. 129-140, 1985.

R. Szeliski. *Bayesian Model of Uncertainty in Low-Level Vision*. PhD thesis, Dept. of Computer Science, Carnegie Mellon Univ., Pittsburgh, 1988.

G.B. Thomas. *Calculus and Analytical Geometry*. Addison-Wesley, Reading, Mass., 1968.

W.B. Thompson and S.T. Barnard. Lower level estimation and interpretation of visual motion. *Computer*, 20(8):20-28, 1987.

C. Tomasi and T. Kanade. Shape and motion without depth. In *Proc. Third Int'l Conf. Computer Vision*, Osaka, Japan, pp. 90-95, 1990.

R.Y. Tsai. Multiframe image point matching and 3D surface reconstruction. *IEEE Trans. Pattern Analysis and Machine Intelligence*, PAMI-5:159-174, 1983.

R.Y. Tsai and T.S. Huang. Uniqueness and estimation of three-dimensional motion parameters of rigid objects with curved surfaces. *IEEE Trans. Pattern Analysis and Machine Intelligence*, PAMI-6:13-27, 1984.

J.K. Tsotsos, D.J. Fleet, and A.D. Jepson. Towards a theory for motion understanding in man and machine. In W.N. Martin and J.K. Aggarwal, eds., *Motion Understanding: Robot and Human Vision*, pp. 353-418. Kluwer Academic Publishers, Boston, 1988.

J.J. Tuma. *Engineering Mathematics Handbook*. McGraw-Hill, New York, 1979.

S. Ullman. *The Interpretation of Visual Motion*. MIT Press, Cambridge, Mass., 1979.

J.P.H. Van Santen and G. Sperling. Elaborated Reichardt detectors. *J. Optical Soc. America*, 2(2):300-321, 1985.

A. Verri and T. Poggio. Against quantitative optical flow. In *Proc. First Int'l Conf. Computer Vision*, London, pp. 171-180, 1987.

H. Wallach and D.H. O'Connell. The kinetic depth effect. *J. Experimental Psychology*, 45:205-207, 1953.

A.B. Watson and A.J. Ahumada. A look at motion in frequency domain. NASA Tech. Memorandum 84352, NASA Ames Research Center, Moffett Field, Calif., 1983.

A.B. Watson and A.J. Ahumada. Model of human visual motion sensing. *J. Optical Soc. America*, 2:322-341, 1985.

A.M. Waxman. Image flow theory: A framework for 3D inference from time varying imagery. In C. Brown, ed., *Advances in Computer Vision*, pp. 165-223. Lawrence Erlbaum Associates, Publishers, Hillsdale, N.J., 1988.

A.M. Waxman, B. Kamgar-Parsi, and M. Subbrao. Closed form solutions to image-flow equations. *Int'l J. Computer Vision*, 1:239-258, 1987.

A.M. Waxman and S. Ullman. Surface structure and three dimensional motion from image-flow kinematics. *Int'l J. Robotics*, 4:72-94, 1985.

A.M. Waxman and K. Wohn. Contour evolution, neighborhood deformation and global image flow: Planar surfaces in motion. *Int'l J. Robotics*, 4:95-108, 1985.

A.M. Waxman, J. Wu, and F. Bergholm. Convected activation profiles and measurement of visual motion. In *Proc. IEEE CVPR*, Ann Arbor, Mich., pp. 717-722, 1988.

A.M. Waxman, J. Wu, and M. Siebert. Computing visual motion in the short and the long: From receptive fields to neural networks. In *Proc. IEEE Workshop on Visual Motion*, Irvine, Calif., 1989.

J. Weng, T.S. Huang, and N. Ahuja. Motion and structure from two perspective views: Algorithms and error analysis. *IEEE Trans. Pattern Analysis and Machine Intelligence*, PAMI-11:451-476, 1989.

T.D. Williams. Depth from camera motion in a real-world scene. *IEEE Trans. Pattern Analysis and Machine Intelligence*, PAMI-2:511-516, 1980.

K. Wohn, L.S. Davis, and P. Thrift. Motion estimation based on multiple local constraints and non-linear smoothing. *Pattern Recognition*, 16:563-570, 1983.

G. Wolberg. *Digital Image Warping*. IEEE Computer Society Press, Los Alamitos, Calif., 1990.

R.Y Wong and E.L. Hall. Sequential hierarchical scene matching. *IEEE Trans. Computers*, 27(4):359-366, 1978.

M. Yachida. Determining velocity maps by spatiotemporal neighborhoods from image sequences. *Computer Vision, Graphics and Image Processing*, 21:262-279, 1983.

M. Ziegler. Hierarchical motion estimation using the phase correlation method in 140mb/s HDTV coding. In *Proc. Third Int'l Workshop on HDTV*, Torino, Italy, 1989.

Author Profile

Ajit Singh was born on October 30, 1963 in Aligarh, India. He received his Bachelor's degree in Electrical Engineering from Banaras Hindu University, Varanasi, India in 1985, a Master's degree in Computer Engineering from Syracuse University, NY, in 1986, and Master's and Ph.D. degrees in Computer Science from Columbia University, NY, in 1988 and 1990, respectively.

He is a scientist in the area of image processing and computer vision at Siemens Corporate Research in Princeton, NJ, and is also an adjunct faculty member at Columbia University. From 1987 to 1989, he worked as a Member of Research Staff in the area of robotics and machine perception at Philips Laboratories in NY. His research interests include image processing, computer vision, robotics, medical imaging and parallel processing and he has published extensively in these areas.

Dr. Singh was the recipient of the University Gold Medal at Banaras Hindu University and Graduate Fellowships at Syracuse University and Columbia University. He is a member of IEEE and ACM, and serves on the editorial board of the IEEE Computer Society Press.

Offices of the IEEE Computer Society

Headquarters Office
1730 Massachusetts Avenue, N.W.
Washington, DC 20036-1903
Phone: (202) 371-0101 — Fax: (202) 728-9614

Publications Office
P.O. Box 3014
10662 Los Vaqueros Circle
Los Alamitos, CA 90720-1264
Membership and General Information: (714) 821-8380
Publication Orders: (800) 272-6657 — Fax: (714) 821-4010

European Office
13, avenue de l'Aquilon
B-1200 Brussels, BELGIUM
Phone: 32-2-770-21-98 — Fax: 32-3-770-85-05

Asian Office
Ooshima Building
2-19-1 Minami-Aoyama, Minato-ku
Tokyo 107, JAPAN
Phone: 81-3-408-3118 — Fax: 81-3-408-3553

IEEE Computer Society

IEEE Computer Society Press Publications

Monographs: A monograph is an authored book consisting of 100-percent original material.

Tutorials: A tutorial is a collection of original materials prepared by the editors, and reprints of the best articles published in a subject area. Tutorials must contain at least five percent of original material (although we recommend 15 to 20 percent of original material).

Reprint collections: A reprint collection contains reprints (divided into sections) with a preface, table of contents, and section introductions discussing the reprints and why they were selected. Collections contain less than five percent of original material.

Technology series: Each technology series is a brief reprint collection — approximately 126-136 pages and containing 12 to 13 papers, each paper focusing on a subset of a specific discipline, such as networks, architecture, software, or robotics.

Submission of proposals: For guidelines on preparing CS Press books, write the Editorial Director, IEEE Computer Society Press, PO Box 3014, 10662 Los Vaqueros Circle, Los Alamitos, CA 90720-1264, or telephone (714) 821-8380.

Purpose

The IEEE Computer Society advances the theory and practice of computer science and engineering, promotes the exchange of technical information among 100,000 members worldwide, and provides a wide range of services to members and nonmembers.

Membership

All members receive the acclaimed monthly magazine *Computer*, discounts, and opportunities to serve (all activities are led by volunteer members). Membership is open to all IEEE members, affiliate society members, and others seriously interested in the computer field.

Publications and Activities

***Computer* magazine:** An authoritative, easy-to-read magazine containing tutorials and in-depth articles on topics across the computer field, plus news, conference reports, book reviews, calendars, calls for papers, interviews, and new products.

Periodicals: The society publishes six magazines and five research transactions. For more details, refer to our membership application or request information as noted above.

Conference proceedings, tutorial texts, and standards documents: The IEEE Computer Society Press publishes more than 100 titles every year.

Standards working groups: Over 100 of these groups produce IEEE standards used throughout the industrial world.

Technical committees: Over 30 TCs publish newsletters, provide interaction with peers in specialty areas, and directly influence standards, conferences, and education.

Conferences/Education: The society holds about 100 conferences each year and sponsors many educational activities, including computing science accreditation.

Chapters: Regular and student chapters worldwide provide the opportunity to interact with colleagues, hear technical experts, and serve the local professional community.